The Neural Control of Sleep and Waking

Springer
New York
Berlin
Heidelberg
Hong Kong
London
Milan
Paris
Tokyo

The Neural Control of Sleep and Waking

Jerome Siegel

With a Foreword by Jerome M. Siegel

With 34 Illustrations

Springer

Jerome Siegel
Departments of Psychology and Biological Sciences
University of Delaware
Newark, DE 19716
USA
jsiegel@udel.edu

Library of Congress Cataloging-in-Publication Data
Siegel, Jerome H., 1932–
    The neural control of sleep and waking / Jerome Siegel.
       p. cm.
    Includes bibliographical references and index.
    ISBN 0-387-95536-4 (hardcover: alk. paper)
    ISBN 0-387-95492-9 (softcover: alk. paper)
    1. Sleep—Research—History.   2. Sleep–wake cycle—Research—History.   I. Title.

QP425 .S585 2002
612.8'21—dc21   2002067019

ISBN 0-387-95536-4 (hardcover)   Printed on acid-free paper.
ISBN 0-387-95492-9 (softcover)

Printed in the United States of America.

9 8 7 6 5 4 3 2 1 SPIN 10884854 (hardcover)   SPIN   10877289 (softcover)

www.springer-ny.com

Springer-Verlag   New York Berlin Heidelberg
A member of BertelsmannSpringer Science+Business Media GmbH

I dedicate this book to my wife, Priscilla Wishnick Siegel. To her I owe a special debt of gratitude for her enormous patience and encouragement over the four years it took to write the book. I dedicate the book also to our three children, Peter, Bennett, and Becky, who periodically remind me that they spent considerable time during their formative years visiting my lab and observing cat surgeries, brain recordings, brain histology, and other aspects of my early research on sleep and waking. I hope this book will put those experiences into a larger and meaningful perspective.

Foreword

My first contact with "the other" Jerome Siegel came in 1973, when I moved to Los Angeles to do postdoctoral work at UCLA. My thesis work had been listed in a nationally available posting without any address. The Brain Information Service, thinking they knew where I was, listed "the other" Jerome Siegel's Delaware address for reprint requests. I soon received a letter from Jerry along with the requests he had received and we have remained in contact ever since. I am occasionally reminded of my namesake when I meet a new colleague who is impressed that someone "so young" published a paper in Science in 1965 (one year out of high school, if it had been me). I entered the field in the early 1970s just as he left. My interests in REM sleep and brainstem mechanisms have been eerily similar to his (and he also did post-doctoral work at UCLA), so our research contributions can be distinguished easily only by my use of my middle initial (which has occasionally been omitted from my publications).

So, my namesake and I both have an interest in seeing to it that no one "brings shame to the name." The current work certainly fulfills that dictum. This is a very unusual book, both in its scope and in its approach to the material. It takes us back to the roots of sleep research going from Galvni's work in 1791, Berger's observations of the electroencephalogram, von Economo's prescient observations growing out of the *encephalitis lethargicia* epidemic at the end of World War I, Kleitman's scholarly work and codiscovery with Aserinsky of REM sleep, and many other important historical developments. He discusses this work clearly and compellingly, presenting not just dates and names, but rather the motivations and findings of these pioneers. The author extracts the enduring scientific and technical significance of this early work in a way that only a sleep scientist could do. The historic context of the early sleep work is explained. Several of the pioneers of sleep research were caught up in the turmoil of World War II. The Nazi-driven deaths of Beck and Berger are rarely presented in histories of the science of brain wave recording. The interruptions caused by the war and the rapid technological advances that followed it are explained.

The book does not talk down to the reader, but rather provides necessary background material so that all can follow the main points of the narrative. When we reach the contemporary issues of sleep science, the work is presented clearly in a way that is accessible to an interested undergraduate. At the same time, frequent "I didn't know that" insights are regularly dished up for experienced sleep researchers such as myself.

The author presents our current understanding of the sleep disorders that afflict more than 10% of the population. The mechanisms, as well as accepted treatments of these disorders are explained in a way that will be useful to those suffering from sleep disruption. Recent advances in our understanding of sleep mechanisms and the 24-hour (circadian) rhythms are presented along with a clear review of current controversies in the field. The recent advances in our understanding of narcolepsy are presented.

The relatively short length of the book was achieved not by omission of key findings. Rather it was accomplished by the author's hard work in clarifying and understanding issues that were occasionally lost on the original authors and presenting them in a very approachable manner. New students of sleep research as well as longtime practitioners will be enlightened by this outstanding book.

Jerome M. Siegel
Professor of Psychiatry and Biobehavioral Sciences
Center for Sleep Research
University of California, Los Angeles

Preface

"Thy best of rest is sleep."
William Shakespeare, *Measure for Measure*, Act 3, Scene 1

Shakespeare succinctly tells us an important truth. After a day's activity, a night spent in bed resting but not sleeping does not refresh us. The next day we are tired and sleepy and often out of sorts. A more serious consequence is that we also are prone to accidents. Daytime sleepiness is a major cause of highway and workplace accidents.

I must admit at the very beginning of the book that how sleep has its beneficial effects and why we need to sleep are not known. Sleep workers have speculated about those issues, and some of their ideas are described in Chapter 13, Theories of Sleep and Waking. Much of the speculation has been informed by the knowledge gained from the biological research on sleep and waking gathered during the past 50 years. We still do not know why we sleep, but we know a fair amount about how we sleep and awaken; that is, we have learned much about the neural mechanisms that control sleep and waking. Most of this book is devoted to describing that work.

The modern era of biological research on sleep and waking began at the turn of the twentieth century. This book surveys that research and briefly includes nineteenth century work upon which the early twentieth century developments were built. The early work, until World War II at midcentury, occurred primarily in Europe. During the first half of the twentieth century, developments in technology and brain research laid the groundwork for the science of sleep and waking. The benefits of this groundwork were reaped during the second half-century. After the war, spurred by the Cold War challenge of the Soviet Union, the United States committed significant resources to research that led to the burgeoning of technological and scientific advances, including many in brain research. Although U.S. science was dominant during the postwar period, research became increasingly international as the countries of Europe and Asia recovered from the devastation of the war.

This book does not provide an exhaustive review of sleep–waking research in the twentieth century. Rather, it describes the major advances in the field, as seen through the eyes of this observer. Some readers may take exception to what I emphasize and what I leave out. This apology has been stated before, most elegantly, by Sir Michael Foster, the founder and first head of the Physiology Department at Cambridge University. In the preface to his book *Lectures on the History of Physiology During the Sixteenth, Seventeenth and Eighteenth Centuries*, published in 1901, Foster writes: "I do not pretend to have given a complete history of physiology even within the period to which I have limited myself. I have chosen certain themes which seemed to me important and striking, and I have striven to develop these, leaving untold a great deal which might be told concerning other themes" (Foster, 1901). From the perspective of a century later, there are obvious gaps in Foster's history. I expect it will not take anything like a century to point out gaps in my presentation here.

I have tried to make the research on sleep and waking accessible to readers who have a background no more extensive than an introductory-level course in biology or psychology in which the basic facts about the nervous system are covered. The book has been written for those who would like a survey of the scientific developments, including the current state of knowledge, in this interesting area of research. The book is also suitable as a text for courses in psychology, biology, or physiology and in the allied health fields. By selecting what I consider to be the major findings in this field, I have kept the book short. My goal has been to provide readers with a brief and manageable overview of the scientific work on sleep and waking. For those interested in more detail and depth, the references provided in the bibliography would be useful.

As an aid to readers not familiar with the technical terms used in this field, a glossary is provided at the end of the book. A list of abbreviations used in this field is also presented. My experience with students who have had an introductory exposure to the nervous system is that they remember the basics about information transfer in the nervous system, such as facts relating to spike action potentials, synaptic connections, and neurotransmitters. However, students are somewhat rusty when it comes to the relative locations of the major structures within the nervous system. I have, therefore, included a short chapter at the beginning of the book that reviews anatomical terms and the basic anatomical organization of the brain—paying more attention to structures that are relevant to sleep and waking.

The subject matter of a book on sleep and waking could be presented in a variety of ways. As is often the case in science, research developments are built upon previous findings and conceptualizations, either as direct follow-ups or as reactions to earlier work. I have, therefore, chosen a chronological approach as a logical and meaningful framework for this material. In addition, the scientific facts and principles are not presented in a social vacuum.

The reader will learn to appreciate that people from far-flung places like Bologna, Italy; Jena, Germany; and Krakow, Poland, made important contributions to an area of current interest. This adds a humanizing dimension to the scientific story.

With respect to research on sleep and waking, the twentieth century divides neatly in half. Research during the first half-century provided the foundation for the major developments that occurred after World War II. However, as can be seen from the list of the contents of the book, the amount of space devoted to each of the two half-centuries is uneven. Part I, The First Half-Century, on historical developments, is considerably shorter than Part II, The Second Half-Century, in which the major advances made after World War II are described. The early groundwork for the postwar advances was certainly important; indeed, the advances in research in the second half-century could not have occurred without the instrumentation developed in the first 50 years. Nevertheless, however important these and the other first half-century contributions were, they do not approach the huge amount of work and the advances in knowledge that followed World War II.

Similarly, in Part II more space is devoted to research on sleep and dreaming (Chapters 6–12) than to the waking part of the story (Chapters 4 and 5). The first major development that ushered in the postwar, modern era of research in this field was published in 1949. This landmark paper was on the brain mechanisms controlling arousal. However, by 1960, research interest had shifted to sleep, where, at the close of the twentieth century and the beginning of the twenty-first century, it remains. Consequently, the chapters on sleeping and dreaming occupy the major share of the book and describe the current state of knowledge in the field.

The last section of the book, Part III, describes the functions and disorders of sleep and waking. Chapter 13 presents a number of theories on the functions of sleep. This chapter is divided into three sections: theories related to the ontogenetic and phylogenetic development of sleep and waking, theories based on the effects of sleep deprivation, and finally, an area of intense current interest, the role of sleep in the process of memory storage. Some of the findings in sleep and waking research have led to increased understanding about the disorders of sleep and waking. These research findings, plus an awareness of the societal costs of sleep and waking disorders, have led to a new specialization in clinical medicine, that of sleep medicine. Workers in this field diagnose, treat, and do research on disorders of sleep and waking. The final chapter of the book describes some of those disorders.

My own background in this area is based on work I had done in the 1960s and 1970s when I published on the neurophysiology and behavioral aspects of sleep and waking. While my research in subsequent years turned to other topics, I have maintained a keen interest in the area of sleep and waking and have kept up with related developments. In my teaching over the years, I have presented this material as a scientific story that evolved and developed in many parts of the world over this past century. That is how the material is presented in this book.

At about the time I left the area of sleep research, Jerome M. Siegel at the University of California at Los Angeles started his career as a sleep scientist. He has since published extensively in this field. Our names are identical except for his middle initial. When his or my work is cited in the text, the context will usually make it clear who is being referred to. If not, when the UCLA Siegel is cited, his middle initial will be included (J.M. Siegel) and citations to my work will be without a second initial.

Jerome Siegel
University of Delaware
Newark, Delaware

Acknowledgments

I am pleased to acknowledge the help of a number of people in the writing of this book. The manuscript was started during a year's sabbatical leave at Oxford University during which Edmund Rolls most generously provided me with a quiet room and a computer for word processing in his research space. He also was close at hand to discuss early formulations of the book and to read and comment on the first draft of the manuscript. I thank Edmund and other members of the Department of Experimental Psychology for their hospitality during that year. Other readers who provided valuable feedback were David Hopkins at Dalhousie University; Evelyn Satinoff, Marvin Zuckerman, and Lawry Gulick at the University of Delaware; Joel Knispel, a former student now in Baltimore; and my wife, Priscilla Wishnick Siegel. Other readers were students who took my "sleep" seminar and used the book manuscript as a text. They provided valuable comments on the writing style and readability from their perspectives as advanced undergraduate majors in biology, psychology, and neuroscience. As expressed in the dedication statement, I owe a special debt of gratitude to my wife for her great patience and emotional support during the writing of the book.

I gratefully acknowledge the skillful help provided by Margie Barrett, the graphics artist and photographer in the Department of Biology at the University of Delaware. She was instrumental in producing the figures in the book. I also appreciate the assistance of the reference librarians at the University of Delaware in tracking down out-of-print and difficult-to-obtain books and journals. The university library also provided me with a faculty study sequestered in the stacks, with all the resources of the library close at hand but removed from the distractions of my office on campus and my study at home. Most of the book, after the year at Oxford, was written there. In a similar vein, I thank my sister and brother-in-law, Florence and Danny Zager, for the use of their Lake George cottage during summer vacations. In that secluded place in upper New York State, I was able to spend uninterrupted days writing and, looking out at that beautiful lake, deriving the inspiration to do so.

Jerome Siegel

Contents

1

A Brief Synopsis of Neuroanatomy

Since much of the information on the brain control of sleep and waking involves a degree of familiarity with brain anatomy, a brief overview of this material is provided at the outset. The description presented here is the general plan of the vertebrate nervous system, with particular attention to parts of the brain involved in the control of sleep and waking. Those who have a background in neuroanatomy can skip this section.

In early embryological development, the central nervous system is in the form of a neural tube. The head end, which will develop into the brain, has three enlargements. From the front (rostral) to the back (caudal), these are the forebrain, midbrain, and hindbrain. These three divisions of the brain are also referred to by their Greek names, the prosencephalon, mesencephalon, and rhombencephalon, respectively. Continuous with the hindbrain caudally is the portion of the neural tube that will develop into the spinal cord. The center of the neural tube is an open space that contains cerebrospinal fluid (CSF). Figure 1.1 illustrates this early condition of the neural tube as well as the brain in later development. Also shown are the major divisions and sub-divisions of the brain and the principal structures located in each subdivision.

With embryological development, the three primitive brain divisions develop and grow in complexity. As this occurs, the hollow interior of the brain that contains the CSF also grows and develops into the four ventricles of the brain. The locations of the ventricles are shown in Figure 1.1. The hindbrain elaborates into two parts. The caudal portion, the medulla oblongata, is continuous with the spinal cord; the rostral portion of the hindbrain is the pons. Moving forward, the ros-

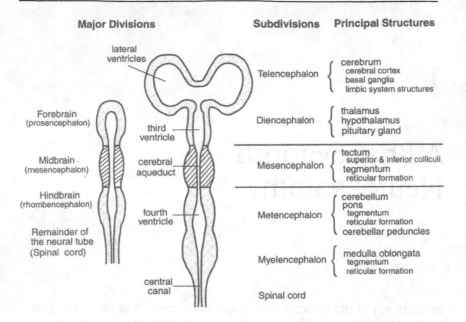

FIGURE 1.1. Schematic representations of the primitive neural tube on the left with the developing fetal brain next to it. The major divisions and subdivisions of the brain are indicated, as are the cerebral ventricles. The principal structures located in different parts of the developed brain are listed on the right.

tral pons is continuous with the midbrain. Unlike the hindbrain and the forebrain, the midbrain (mesencephalon) is not subdivided. The medulla, pons, and mesencephalon comprise the brain stem, which is the basic foundation of all vertebrate brains. Even in the brains of adult animals, these parts of the lower brain still resemble the primitive neural tube in its basic shape. The other parts of the brain, the cerebellum and the forebrain, can be seen as growing out of the brain stem.

The cerebellum emerges from the brain stem at the level of the pons via three pairs of large fiber bundles, the superior, middle, and inferior cerebellar peduncles. The superior and inferior peduncles are the most rostral and caudal pairs, respectively. As the cerebellum develops and grows, it covers the dorsal (top) part of the pons and even the medulla behind the pons.

Most important for the subject of sleep and waking is a structure called the reticular formation, which comprises the central core of the tegmentum. The tegmentum and its reticular core extend throughout the entire length of the brain stem from the caudal medulla to the rostral mesencephalon. Figure 1.2 shows the location of the reticular formation in a schematic view of a cat brain that has been sectioned vertically about 0.5 mm from the midline to separate the brain into left and right halves. This section plane is called a sagittal section. In Figure 1.2 we are looking at the medial aspect of the left half of the brain. This basic drawing, with additional structures labeled, will be used throughout the book to show the brain regions and neural pathways that play a role in the control of sleep and waking.

The reticular formation is a network of neuronal cell bodies interlaced with nerve fibers. Some of the cell bodies are clustered into well-defined nuclei; others are loosely distributed to comprise "fields" of cells. Interspersed among the populations of cell bodies are axons that interconnect the nuclei and cell fields of the reticular formation. There are also long fiber tracts that conduct information through the reticular formation as well as into and out of it. Many of the nuclear groups and fiber connections of the brain stem reticular formation are of central importance in the initiation and regulation of sleep and waking.

All vertebrates, from primitive forms such as amphibians to advanced

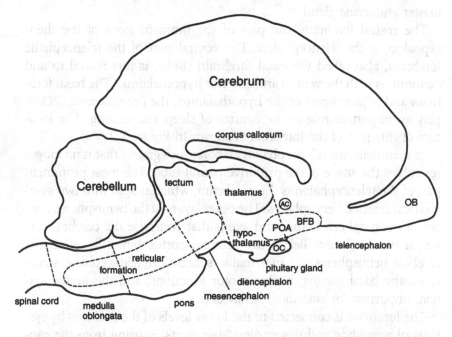

FIGURE 1.2. Drawing of the adult cat brain based on a sagittal section about 0.5 mm lateral to the midline. This view shows the location of the reticular formation of the brain stem relative to other structures of the brain. Also represented are the boundaries (dashed lines) that demarcate the five subdivisions of the brain: the medulla oblongata and pons (the hindbrain), the mesencephalon (midbrain), and the diencephalon and telencephalon (the forebrain). Abbreviations: AC, anterior commissure; BFB, basal forebrain; OB, olfactory bulb; OC, optic chiasm; POA, preoptic area.

mammalian species, have brain stems similar in design but of different absolute sizes. The brain differences among species become dramatic when we look at the elaboration of the forebrain that is the enlarged rostral part of the neural tube. The great growth of the rostral end of the neural tube to form the phylogenetically newer parts of the brain is a process known as encephalization. In advanced species, such as mammals, the rostral part of the forebrain develops into a very complex structure that completely covers and dwarfs the brain stem. In contrast, the forebrain is relatively small in lower vertebrate forms.

The forebrain, like the hindbrain, elaborates into two parts (refer to Figures 1.1 and 1.2). The caudal portion, adjacent to and just in front of (rostral to) the midbrain, is the diencephalon. Some neuroanatomists prefer to classify the diencephalic portion of the forebrain as part of the brain stem. The main components of the diencephalon are the thalamus, located in the dorsal part of the diencephalon, and the hypothalamus, which is below (ventral to) the thalamus. The ventral hypothalamus is connected via the pituitary stalk to the pituitary gland, the master endocrine gland.

The rostral forebrain, the part of the brain in front of the diencephalon, is the telencephalon. The ventral part of the telencephalic forebrain, also called the basal forebrain (BFB), is just rostral to and continuous with the ventral and anterior hypothalamus. The basal forebrain and adjacent part of the hypothalamus, the preoptic area (POA), play an important role in the control of sleep and waking. The location of this part of the forebrain is shown in Figure 1.2.

In mammals, the telencephalon elaborates so greatly that it no longer resembles the shape of the primitive neural tube. The most prominent part of the telencephalon is the cerebrum, which is split into two symmetrical cerebral hemispheres. The outer layers of the hemispheres consist of neuronal cell bodies and axons that comprise the cerebral cortex of the cerebrum. Below the cerebral cortex but still within the cerebral hemispheres are two major subcortical telencephalic structures: the basal ganglia, having motor functions, and the limbic system, important for emotional behaviors.

The forebrain is connected to the lower levels of the neuraxis by systems of ascending and descending fiber tracts. Starting from the caudal end of the neuraxis and progressing to the front, the spinal cord contains ascending sensory fiber tracts that carry touch, temperature, and pain messages from receptors located throughout the body. These fibers ascend the spinal cord and course through the brain stem toward the sensory nuclei of the thalamus. In addition, proprioceptive sensory signals that originate in muscles of the body also ascend the spinal cord and enter the brain stem, but they travel to the cerebellum via a fiber bundle, the inferior cerebellar peduncle. The proprioceptive information is processed by cerebellar circuits and conducted back to the brain stem via the superior cerebellar peduncle to continue up to the thalamus. Sensory messages that originate in receptor systems located in the head (e.g., visual and auditory) enter the brain stem via cranial nerves and also send their signals to the thalamus. The sensory nuclei of the thalamus as well as all other thalamic nuclei (barring one, the reticular nucleus) send their projections to the cerebral cortex as components in a very large fiber bundle, the internal capsule. The internal capsule contains, in addition to the thalamocortical axons, all other subcortical projections to the cortex. Similarly, all descending fibers from the cortex to lower parts of the nervous system are carried in the internal capsule.

The sensory nuclei of the thalamus project to specific sensory receiving areas of the cortex. Another group of thalamic nuclei project to broad areas of cortex and have been described as diffuse or nonspecific projection nuclei. They include the intralaminar nuclei of the thalamus. These nuclei play an important role in determining the sleep and waking states of the cerebral cortex by virtue of their projections to widespread areas of the cortex.

At one time, it was thought that the thalamus was the sole gateway to the cortex; that is, all information to the cortex was believed to be funneled through the thalamus. With the advent of sensitive neuroanatomical tracing and histochemical techniques, it was discovered that there are other projections to the cortex in addition to those from the thalamus. These other projections to the cortex originate in several nuclei located in the brain stem and are sometimes referred to as extrathalamic cortical projections. These subcortical nuclei project fibers that release known neurotransmitters (dopamine, noradrenaline, serotonin, histamine, and acetylcholine) to broad areas of cortex. By virtue of their diffuse projection patterns, these nuclei are similar to those of the intralaminar nuclei. The various transmitters they release serve to modulate cortical functions—in some cases, they alter attention and arousal levels.

In addition to the ascending connections, the forebrain is connected to the brain stem via descending connections. The efferent fibers that originate in the motor cortex and related areas of the frontal cortex course toward the brain stem as a large fiber bundle, the pyramidal tract. The pyramidal tract comprises a substantial portion of the fibers within the internal capsule and enters the brain stem at the base of the mesencephalon. Motor fibers controlling muscles of the head and face leave the pyramidal tract in the brain stem to connect with motor nuclei of cranial nerves—for example, the oculomotor nerve to control eye movements. The motor fibers that control muscles of the body continue through the brain stem as the corticospinal component of the pyramidal tract and connect to spinal motor neurons of spinal nerves.

Finally, often overlooked are cortical efferent nonmotor fibers that project from the cerebral cortex to the thalamus. Areas of the cortex that receive inputs from thalamic nuclei reciprocate by feeding back to those nuclei corticothalamic fibers via the internal capsule. These thalamocortical and corticothalamic loops play a role in regulating the arousal level of the cortex.

A comment about terminology. The terms afferent and efferent are used frequently in this book, and their meanings are sometimes misunderstood. "Afferent" is often used in conjunction with sensory pathways and "efferent" with motor pathways. However, the more basic meanings of these words in neuroanatomy are with reference to fiber tracts that project *from* a given location or *to* a location—independent of sensory or motor functions. For example, *with respect to nucleus A*, fibers that originate in A and project from there are efferent fibers. If these fibers project to nucleus B, *with respect to B*, they are afferent fibers. Fibers that leave a nucleus are efferent with reference to that nucleus; fibers that come into a nucleus are afferent with respect to that nucleus.

Part I
The First Half-Century: The Groundwork for the Science of Sleep and Waking Is Laid

2

Technological Developments

Electroencephalography: Hans Berger

The first major advance in the area of sleep and waking in the twentieth century was the development of technology that could record the electrical activity of the brain. The instrument or machine that does this is the electroencephalograph and the record of electrical activity produced by the machine is the electroencephalogram. Both the machine and the record are abbreviated as EEG. So we often see terms like "EEG machine," "EEG record," and "EEG signals." This technology provided the single most important tool that enabled subsequent advances to be made. It is impossible to imagine how our current knowledge about the functioning of the brain could have been acquired without our ability to record its electrical signals. The introduction and first application of EEG technology is usually attributed to Hans Berger (1873–1941), a professor of neurology and psychiatry at the Friedrich Schiller University in Jena, Germany. Berger's discovery of the EEG in humans was built upon the work of four scientists at the end of the eighteenth century and in the nineteenth century. These scientists, an Italian, a German, an Englishman, and a Pole, did the pioneering work on animals that demonstrated neural activity was accompanied by detectable electrical changes.

Of the four historical figures, Luigi Galvani (1737–1798) made the earliest contribution. In the *Proceedings of the Bologna Academy* in 1791, Galvani, professor of anatomy at the University of Bologna, claimed that neural tissue, in this case the sciatic nerve of a frog nerve–muscle preparation, was capable of generating an electrical signal when active. Moreover, he showed that the nerve had conductive properties that permitted it to carry the electrical signal to muscle and thus cause a

9

contraction. At the time, Galvani was engaged in a heated controversy with Alessandro Volta (1745–1827), a professor of physics at the University of Pavia, over whether nerves were not only conductive (this was not contested), but generative of electrical energy as well. Though Galvani was not strictly accurate in all his interpretations, in the final analysis, his position was correct. Neural tissue has the intrinsic property of generating an electrical signal. Further developments in electrophysiology rapidly followed Galvani's demonstration.

In the mid-nineteenth century, important advances in physiology were made in Germany by Johannes Muller and his students and colleagues in Berlin, one of whom was Emil Du Bois-Reymond (1818–1896). Du Bois-Reymond used a frog nerve–muscle preparation, similar to that used by Galvani, and an induction coil to electrically activate the peripheral nerve. However, instead of merely observing the muscle twitch to infer that an electrical nerve signal had occurred, Du Bois-Reymond, in 1849, used a galvanometer, a device invented by the Danish scientist Hans Christian Oersted just a few years earlier, in 1840, to directly monitor the electrical event. Du Bois-Reymond detected a negative electrical transient change in the nerve induced by the stimulation. He directly observed what Galvani, years earlier, had postulated was the signal energy in nerve. His was the first detection of the compound action potential in a peripheral nerve.* In her excellent book, *A History of the Electrical Activity of the Brain,* Mary Brazier, an electrophysiologist with an interest in the history of science, comments that "Du Bois-Reymond, himself, did not underestimate the importance of his discovery" (Brazier, 1961, p. 2). She presents a statement from Du Bois-Reymond's publication of 1849:

> If I do not greatly deceive myself, I have succeeded in realizing in full actuality (albeit under a slightly different aspect) the hundred years' dream of physicists and physiologists, to wit, the identification of the nervous principle with electricity (quoted in Brazier, 1961, p. 2).

What Du Bois-Reymond had demonstrated in a peripheral nerve in 1849 was first shown in the brain by Richard Caton (1842–1926) in

---

*The compound action potential in a peripheral nerve is a reflection of the firing of axons in a whole nerve bundle. When a nerve is activated by a discrete stimulus, the individual axons of the fiber bundle will fire their action potentials in near simultaneity. The neural activity detected from the whole bundle is, therefore, referred to as a *compound* action potential.

1875 at the University of Liverpool in England. Caton, using a more sensitive galvanometer than that used by Du Bois-Reymond, was able to record voltages of varying frequencies and low amplitudes from the cortical surface of the rabbit and monkey brain (Caton, 1875). In the 1875 and subsequent publications, Caton showed that sensory stimulation produces electrical responses in specific areas of the cerebral cortex. In these papers he also reported changes in cortical rhythms as a function of sleep and waking.

Such recordings from the cortical surface of the brain are not quite the same as the compound action potentials recorded from a bundle of nerve fibers. Cortical recordings are due to graded potentials generated by the large dendritic fields of cortical neurons that exist close to the surface of the cortex. Graded field potentials are voltages of varying amplitudes that are in contrast to the "all-or-none" action potentials generated by axons. At any point in time these field potential recordings reflect the algebraic summation of the graded potentials that are in the vicinity of the recording electrode. If the cortical activity is not related to the occurrence of a sensory stimulus or an event, it is referred to as "spontaneous" cortical activity. However, if the cortical activity is generated by a discrete stimulus (i.e., if its occurrence is time-locked to the presentation of a sensory or an electrical stimulus), it is referred to as an "evoked" field potential, which is somewhat comparable to an evoked compound action potential. Field potential recordings that are not time-locked to stimuli comprise the electrical rhythms we now know as the electroencephalogram (EEG). Caton was the first to report these brain wave rhythms from the cortical surface of animals.

A few years after Caton published his early papers, Adolf Beck (1863–1942), at the University of Krakow in Poland, unaware of Caton's work, independently arrived at similar findings. For his doctoral thesis published in 1891, Beck, working with rabbits and dogs, observed different brain rhythms related to the state of an animal. He also reported, for the first time, that the typical high voltage brain wave pattern seen in sleep could be blocked by sensory stimulation of any sort (Beck, 1891).

In 1895 Beck moved to the University of Lvov, then in eastern Poland, now a part of Ukraine. Beck, as described by Brazier (1961), had a long and illustrious career at Lvov, rising to positions of dean and rector at his university. On September 1, 1939, Germany invaded Poland, and by 1942, with the German program of the "final solution" for the Jews of Europe well under way, Beck tragically committed suicide at his university hospital. Brazier movingly describes the events in Lvov:

> As the Germans closed in on Lvov the danger to Beck increased, for he was Jewish. An old man now, rather than go into hiding, he chose to stay in the shadow of the University to which he had given so many years of his life. Just before his eightieth birthday, he became unwell and while he was in the hospital for an ailment, the Germans came to take him to the extermination camp. Beck's son, a physician, had supplied all members of the family with capsules of potassium cyanide. Beck took his capsule and saved himself from the gas chamber (Brazier, 1961, p. 48).

Caton and Beck had shown various types of rhythmic electrical oscillations from the brains of animals, but it remained to be demonstrated that the human brain also generated electrical rhythms. Hans Berger in Jena, Germany, was familiar with Caton's reports and over a period of years prior to and after World War I he attempted to extend Caton's work by recording evoked electrical responses in the animal brain to a variety of sensory stimuli. However, the string galvanometer he used was not sufficiently sensitive to accomplish this. In 1924 Berger turned to recording from the human brain, using electrodes placed on the surface of the scalp or, in some cases, fine needlelike wires inserted into the scalp. Detecting voltages from the scalp is more demanding than recording from the exposed brain of an animal. This is because the voltages, already as low as microvolts, are attenuated by approximately 50% when the resistance of the skull and scalp are interposed between the voltage sources in the brain and the recording electrodes. In 1925 Berger had access to a state-of-the-art galvanometer manufactured by Siemens, a German company specializing then and to this day in the development and manufacture of electrical instruments. Berger acquired this sensitive galvanometer after his daughter married an engineer who worked for Siemens. The first electroencephalographic record from a human was made in 1925 and published four years later in 1929 (Berger, 1929). The subject for this historic brain recording was Berger's own young son. In a series of 23 papers published between 1929 and 1938, Berger reported EEG responses in humans evoked by a variety of sensory stimuli, including painful ones, and accurately described the different EEG rhythms seen during sleep and waking.

In 1990, at the time of the collapse of the Soviet Union and the fall of the Berlin Wall, I visited the German Democratic Republic (the GDR), better known as East Germany, to give a talk in the Department of Physiology at the Friedrich Schiller University, in Jena. On the walls of the hallway and stairwell were photographs and portraits of previous distinguished faculty, among which was a prominently displayed picture of Hans Berger. I asked about him, and the head of the department, Professor Wolfgang Haschke, now retired, had a wealth of information. He described Berger as being reclusive and secretive about his research. Berger would seclude himself in his laboratory and very few knew what he was doing. He apparently worked on animal and later human brain wave recording for over 10 years before publishing his findings. Berger, like Beck, rose to the position of rector of his university and tragically, also like Beck, committed suicide in his own university hospital. In 1938, Berger was forced to vacate his position at the university owing to his anti-Nazi sentiments. His laboratory was dismantled, and he experienced a prolonged period of depression. In 1941 he entered the hospital as a patient and took his own life by hanging. Figure 2.1 is

FIGURE 2.1. Photographic portrait of Hans Berger. (Reproduced with permission from K. Kolle, Grosse Nervenarzte, 1956, facing page 4. Stuttgart: Georg Thieme Verlag.)

a photographic portrait of Hans Berger. Interesting accounts of Berger and his EEG work are provided by Brazier (1961, 1980) and Jones (1995).

The EEG in Sleep and in Waking

Because EEG recordings have been so central to sleep–waking research and are referred to frequently in this book, a description of the EEG in sleep and in waking is in order. Humans and other mammals show basic similarities in their sleep–waking records, with the human tracings being somewhat more complex.

Soon after objective measures of dream sleep were discovered, it became conventional to categorize sleep as either dream or nondream sleep. Since dreaming is objectively measured by the occurrence of rapid eye movements (REMs), dream sleep is usually called REM sleep and nondream sleep is known as non-REM (NREM) sleep. Because EEG records during NREM sleep show recognizable and consistent changes during the course of sleep, NREM sleep has been differentiated into four stages based on these changes.

Figure 2.2A shows examples of EEG tracings recorded during waking and the stages of sleep in a human. Figure 2.2B displays a derived

───▶

FIGURE 2.2. (A) Typical EEG recordings during waking and the stages of sleep. The top trace, recorded during alert waking, is characterized by low voltage beta waves in the frequency range of 20 to 40 Hz. The trace marked "Awake with eyes closed" shows the predominantly 10 Hz alpha rhythm. Stage 1 non-REM sleep is a short period between waking and deeper states of sleep that shows some beta waves with slower frequencies interspersed. Sleep stages 2 and 3 show a mixture of beta-like waves that are remnants of stage 1 sleep mixed with sleep spindles having a frequency of 12 to 15 Hz that are characteristic of sleep stages 2 and 3. During stage 3, delta waves start appearing and dominate the record during stage 4 sleep. The stage 4 trace shows the high voltage, low frequency (0.5–3 Hz) delta waves. The bottom trace, taken during REM sleep, resembles the low voltage, high frequency waveform seen in an awake and aroused cortex. During REM sleep, the individual is not awake, but the cortex appears aroused. (Reproduced with permission from D. Purves, G. J. Augustine, D. Fitzpatrick, L. C. Katz, A.-S. LaMantia, and J. O. McNamara, eds., Neuroscience, 1997, p. 498. Sunderland, MA: Sinauer.) (B) A typical hypnogram, derived from an all-night sleep recording, graphically summarizes the sleep stages an individual experiences during a night's sleep. This hypnogram shows the descent from waking through stages 1, 2, 3, and 4 within 45 minutes of falling asleep, dwelling at stage 4 for 15 minutes and ascending to the REM stage at 90 minutes after falling asleep. The hypnogram shows the cycle repeating about every hour-and-a-half to yield five REM episodes during a night's sleep. It also shows less stage 4 sleep and longer REM sleep periods as the night progresses.

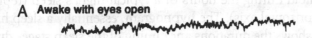

A Awake with eyes open

Awake with eyes closed

Non-REM sleep
Stage 1

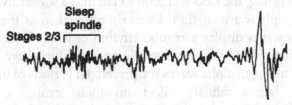

Sleep
spindle

Stages 2/3

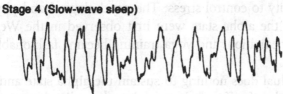

Stage 4 (Slow-wave sleep)

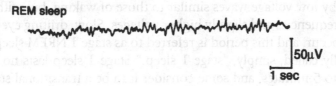

REM sleep

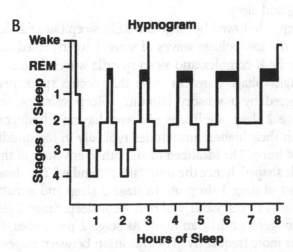

50μV

1 sec

B   **Hypnogram**

Wake

REM

1

2

3

4

1   2   3   4   5   6   7   8

Hours of Sleep

Stages of Sleep

hypnogram that shows the sleep stages (represented on the vertical axis) as they unfold during the hours of a typical night's sleep (represented on the horizontal axis). The hypnogram, essentially a sleep histogram, readily shows the durations and 'iming of the sleep stages during a night's sleep.

During attentive waking the EEG is comprised of low voltage, high frequency (LVHF) waves, also called beta waves. The term low voltage fast activity (LVFA) is also commonly used. Low voltage refers to amplitudes of 5 to 10 microvolts (μV) recorded from the scalp (where human recordings are usually taken); high frequency (or fast activity) refers to rhythms in the range of 20 to 40 hertz (Hz). On the other hand, when a person is awake but relaxed and not thinking about or attending to anything, the EEG will tend to fall into a slower frequency known as the alpha wave rhythm, formerly referred to as the Berger rhythm. Alpha waves display a regular rhythm of 8 to 12 Hz, higher in amplitude than the LVHF beta waves. Some individuals have learned to control and maintain alpha waves for prolonged periods of time and claim that it has a salutary effect on their feelings of well-being and ability to control stress. The psychological and physiological benefits of the alpha state were first observed in the West when biofeedback, meditation, and yoga training became fashionable in the late 1960s.

If an individual does nothing to sustain the alpha state and continues to relax and drift off into sleep, alpha waves disappear and are replaced by low voltage waves similar to those of waking, but with a few lower frequencies and slightly higher voltages. Slow, drifting eye movements occur, and this period is referred to as stage 1 NREM sleep. This is usually called, simply, "stage 1 sleep." Stage 1 sleep lasts no longer than 3 to 5 minutes, and some consider it to be a transitional state between waking and sleep.

Stage 1 sleep is followed by stage 2 NREM sleep (stage 2 sleep) that is marked by the low voltage waves of stage 1 interspersed with what are known as the K-complex and sleep spindle waves. The K-complex is a sharp, high voltage transient wave that occurs spontaneously or may be triggered by a sensory stimulus. Sleep spindles, the other marker of stage 2 sleep, are bursts of waves having a frequency of 12 to 15 Hz, with their highest amplitudes typically in the middle of the 1- to 2-second burst. The idealized or smoothed envelope of this waveform is spindle shaped, hence the term "sleep spindle." The slow, rolling eye movements of stage 1 drop out in stage 2 sleep and remain absent in the subsequent stages 3 and 4 of non-REM sleep. Stage 2 sleep usually lasts no longer than half an hour. As stage 2 progresses, the sleep spindles occur more frequently. The transition between stages 2 and 3 sleep occurs when the low voltage background waves that separate the spindles are replaced by a high amplitude, low frequency rhythm called delta waves. Delta waves are normally the highest amplitude EEG waveforms observed. They may reach amplitudes of about 300 μV. They are also the slowest EEG rhythms seen, namely, 0.5 to 3 Hz. Stage 4 sleep is marked by the proportion of spindles to delta shifting in favor of more delta waves. In deep stage 4 sleep, spindles drop out and the record consists almost solely of large rolling delta waves.

Sleep stages 3 and 4 are referred to as slow-wave sleep (SWS), the slow waves being the sleep spindles and delta waves. Slow-wave sleep is also referred to as high voltage, low frequency (HVLF) sleep. Another widely used term is synchronized sleep, based on the notion that the high voltage is due to the synchronized activity of many neurons during the slow-wave stages of sleep. Similarly, a low voltage, high frequency (LVHF) record seen during waking and dream sleep is referred to as desynchronized sleep or, simply, as desynchronization. The notion here is that the same populations of cortical neurons are now active in many different firing patterns to support the cognitive and perceptual processes that occur during waking and dreaming, but the timing of their firing with respect to each other is desynchronized.

As shown in the hypnogram in Figure 2.2B, the stages of sleep descend from the waking state through stages 1, 2, and 3 of NREM sleep to stage 4, delta-wave sleep, the deepest stage of NREM sleep. In humans, stage 4 sleep begins a little less than an hour after sleep onset and lasts 20 to 30 minutes, after which the delta EEG waves give way to a few minutes of stage 3 sleep followed by stage 2. There is then a rapid transition to REM sleep that occurs, typically, about 90 to 100 minutes into a night's sleep. As also seen in the hypnogram, the transition from stage 4 NREM to REM sleep is represented as an ascending process. Some authors show REM sleep as a period directly following and, on the hypnogram, below that of stage 4 sleep. However, since the transition between stage 4 and REM sleep usually includes a brief excursion through stages 3 and 2, I prefer to indicate the transition into REM sleep, as shown in Figure 2.2, as an ascending process. Also, representing it that way places REM sleep close to the waking state. This is consistent with the observation that cortical activity during REM sleep and waking are similar—waking and dreaming are very similar brain states. The EEG during REM sleep is dominated by low amplitude beta waves that also characterize the waking EEG, but during REM sleep there are now rapid eye movements and the individual is unambiguously asleep.

The ascent from stage 4 NREM to REM sleep is more rapid than the descending process. It takes almost an hour to go through the descending phases from waking to deep slow-wave sleep and only a matter of minutes to ascend from stage 4 sleep back through the lighter stages of NREM sleep to the REM stage of sleep. A number of terms have been used to refer to REM sleep. One of these is D sleep, meaning "dream sleep." However, "D sleep" has also been used to mean "deep sleep," referring to the REM stage. "Deep sleep" is an unfortunate term, because depending upon the procedures and circumstances of measuring arousal thresholds, stage 4 delta sleep may be a deeper sleep than REM sleep. Since the descriptive "D" may refer to "dream," "deep," or "delta," the term "D sleep" is best avoided.

The first REM sleep episode lasts only 5 to 10 minutes. Interestingly, the end of each REM episode is usually punctuated by a large body movement, often a shift in body position, and a very brief awakening. This typical pattern has been observed in humans, cats, and other mammals. After the REM episode, sleep will cycle down again through stages 2, 3, and 4 and back again to REM sleep. The cycle period is about 90 minutes, with successive cycles showing less slow-wave sleep (stages 3 and 4) and longer durations of REM sleep. The last REM sleep episode in the morning may be as long as 30 to 50 minutes, after which the person awakens for the day. The cyclical nature of the sleep stages is clearly seen in the hypnogram of Figure 2.2B.

In recent years, new findings in the EEG field have complicated the convention in which the term "desynchronized EEG" is synonymous with the low voltage, fast activity (LVFA) beta waves (20–40 Hz). A number of labs have shown that during waking and dream sleep a desynchronized EEG is "more apparent than real," as stated by Steriade (1996). What this means is that a more sophisticated analysis of the EEG reveals that during waking and dream sleep, a very low voltage rhythm in the range of 30 to 60 Hz (but mainly 40 Hz) is embedded in the low voltage EEG and that this rhythm represents a synchronization of neuronal activity. The predominantly 40 Hz oscillation is called a "gamma rhythm." Workers in this field present data to support the position that this fast rhythm represents a temporal synchronization and coordination of neuronal activity in anatomically distributed cortical and subcortical circuits that occur during waking and dream sleep (Singer, 1993; Munk et al., 1996; Steriade, 1996; Steriade et al., 1996a,b; Chrobak and Buzsáki, 1998; Miltner et al., 1999; Rodriguez et al., 1999). Their view is that the gamma rhythm reflects a temporal binding of neural activity in widely distributed circuits into functional neural assemblies.

This is a difficult concept to grasp. Think of the perception of a given event. The perception contains many different sensory and complex cognitive components. However, the conscious perception is experienced, essentially, as a unitary phenomenon. Some binding of neural activity must have occurred to accomplish this. The concept of binding refers to the integration of neural activity that occurs in different brain areas during the same period of time. It is considered necessary for the coordination and linkage of multiple sensory messages and other information from the different brain regions to produce the unified perceptions and cognitions that mark our waking behavior. During the cortical arousal that occurs in dream sleep, the synchronized gamma oscillations reflect the binding of neural activity to produce the *internally* generated perceptual and cognitive experiences.

Sleep in lower mammals like cats, although similar to that of humans, is not differentiated into as many stages. It is more simply described as slow-wave sleep (stages 3 and 4 in the human) and REM sleep. In animals, the REM stage is often referred to as the paradoxical phase of sleep, as will be discussed later. The durations of sleep stages and cycle times are shorter than in humans. In animals it is usual to place recording electrodes directly on the cerebral cortex. Because the electrodes are closer to the voltage source, the EEG amplitudes are larger. Voltage amplitudes recorded from cat cortex range from 20 μV beta waves during waking to 500 μV (0.5 mV) delta waves seen during stage 4 sleep. The comparable EEG waves recorded from the human scalp are about half these amplitudes.

Electroencephalography has proven to be very valuable in clinical use as well as in research. For example, certain forms of epilepsy are difficult to diagnose and detect, but their EEG manifestations are clear. The focus of epileptic discharge can be detected by scalp electrodes, and they have been used to localize the area of brain tissue that discharges abnormally. In some cases, even the EEG signs of pathophysiology are elusive and show up only during sleep recordings. Therefore, the EEG service in hospitals often has facilities to do sleep recordings. In recent years, sleep clinics centered around an EEG facility have become common in the effort to diagnose and treat a variety of sleep disorders.

For the first half of the twentieth century, the EEG proved to be a most valuable tool in sleep and waking research, and it continues to be the mainstay of this research. However, between the two world wars other approaches took center stage in yielding information about the biology of sleep and waking.

The Stereotaxic Instrument:
Victor Horsley and Robert Clarke

The stereotaxic instrument, first seeing use in the early 1900s, was another technological development essential to the advancement of research in sleep and waking because it enabled precise and repeatable access to structures deep within the brain. The outer cell layers of the cerebrum, the cerebral cortex, display large changes in electrical activity during sleep and waking. It is apparent from the EEG that the cortex functions very differently in these two states. However, the different electrical rhythms of the cortex are governed not by the cortex itself but by subcortical structures that lie deep within the brain. These deep structures, unlike the cerebral cortex, cannot be experimentally manipulated under direct visual control. The stereotaxic instrument permitted subcortical structures to be accurately probed and manipulated. By means of electrical and chemical stimulation, lesions, and electrical recording, the roles of deep structures in the initiation and maintenance of sleep and waking have come to be understood.

In two articles that appeared in the *Kopf Carrier,* published by David Kopf Instruments, the leading manufacturer of stereotaxic instruments, Louise Marshall, a historian of science at the University of California at Los Angeles (UCLA), and Horace Magoun, a professor of anatomy at UCLA, described the history of the development of this device (Marshall and Magoun, 1990, 1991). There were efforts early in the nineteenth century to develop an instrument to probe subcortical structures of the brain. However, Marshall and Magoun show that in the early years of the twentieth century, Victor Horsley (1857–1916) and Robert Clarke (1850–1926) developed the concept and design of the stereotaxic instrument that is most widely used today. In fact, the instrument is often referred to as the Horsley–Clarke stereotaxic instrument. Figure 2.3 shows a modern stereotaxic instrument and probe drive manufactured by David Kopf.

Clarke and Horsley, working at University College in London, published the first account of their instrument in 1906 (Clarke and Horsley, 1906). The first instrument, constructed by a machinist at the Palmer Company in London, and other versions that followed, were designed for cat, monkey, and man. The Horsley–Clarke stereotaxic instrument is based on rectilinear coordinates in each of the three Cartesian axes, x, y, and z. The head is securely fixed in the instrument in a standard orientation with respect to a probe drive that can be moved along each of the three calibrated axes. Movements of the stereotaxic probe drive correspond to the three standard anatomical spatial dimensions: anterior–posterior (interchangeable with rostral–caudal for the brain), medial–lateral,

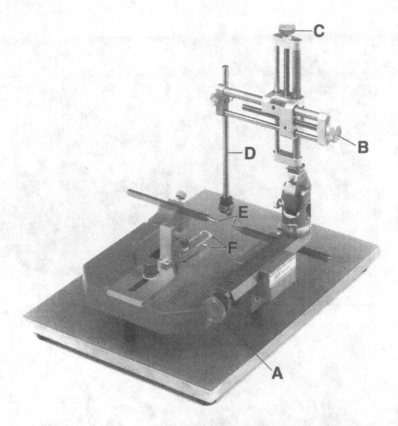

FIGURE 2.3. The stereotaxic instrument. This particular instrument is designed for the rat. The head of an anesthetized rat is securely fixed by the ear bars (E) and the mouth and nose plates (F). The probe drive has an electrode holder, the vertical rod (D) to which the probe is attached at its lower end. The probe drive can move an electrode in each of the three stereotaxic coordinates. Knobs A, B, and C move the probe in the anterior–posterior, medial–lateral, and dorsal–ventral dimensions, respectively. The probe drive is first manipulated by knobs A and B to position the probe above the skull to its anterior–posterior and medial–lateral positions above the target site in the brain. Knob C will then lower the probe tip to its target. (Adapted with permission from David Kopf Instruments Stereotaxic Catalog, 2001, Model 900, p. 6.)

and dorsal–ventral. Each of the three axes is orthogonal to the other two. The anterior–posterior dimension extends from the rostralmost point of the frontal lobe to the back of the brain. The medial–lateral dimension extends from the midline to the lateralmost extent of the brain, both left and right. The dorsal–ventral dimension extends from the top or dorsal aspect of the brain down to its most ventral level. To reach a particular point in the brain, a vertically oriented electrode attached to the probe drive would be moved above the skull to its appropriate anterior–posterior and lateral coordinates and, via the vertical movement of the probe drive, lowered through a hole in the skull into the brain to its target site at the dorsal–ventral (horizontal or depth) coordinate.

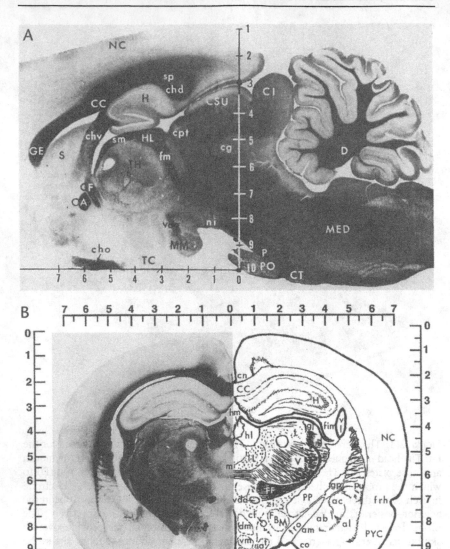

To realize the usefulness of the stereotaxic instrument, brain atlases showing the locations of brain structures with reference to the system of stereotaxic coordinates were required. Brain atlases display series of photographic plates of brain sections that are cut in orientations that correspond to the three axes of the stereotaxic instrument. A transection through the entire brain along the midline produces a midsagittal section, dividing the brain into equal left and right halves. Sections that are parallel and lateral to this are called, simply, sagittal sections. Orthogonal to sagittal sections are horizontal cuts that section the brain at various dorsal–ventral levels. Finally, the most common types of section are cuts through the brain at right angles to each of the other two axes at various anterior–posterior levels. These are called frontal, coronal, or transverse sections or, less formally, cross sections. Most brain atlases show series of frontal sections through the entire brain.

The brain sections are suitably stained to show the clusters of neuronal cell bodies that comprise the nuclei of the brain and the fiber tracts that connect the nuclei. Many atlases also have drawings that accompany the stained brain sections. The drawings are helpful in that they outline the approximate borders between structures that are often difficult to discern in photographs of the stained sections.

Clarke, collaborating with a surgeon/illustrator, E. Erskine Henderson, published the first stereotaxic brain atlas based on this system of coordinates (Clarke and Henderson, 1912). This atlas contained plates of sagittal sections through the cat brain that permitted a probe to be placed in a subcortical nucleus or fiber tract defined by the three stereotaxic coordinates. Since the publication of the Clarke and Henderson atlas of the cat brain in 1912, brain atlases of many vertebrate species, including *Homo sapiens*, have been published. Most of these atlases locate brain structures utilizing the Horsley–Clarke stereotaxic coordinate system. Figure 2.4 shows a sagittal and a frontal section from a stereotaxic atlas of the rat brain (Massopust, 1961).

FIGURE 2.4. Sagittal and frontal (coronal) sections from an atlas of the rat brain. The sagittal section (A) was cut 1.0 mm lateral to the midline to reveal midline structures throughout the anterior–posterior and dorsal–ventral aspects of the entire brain. Since this is an atlas of the rat diencephalon, the anterior–posterior coordinates are given only from A–P zero to anterior 7, the region that includes the diencephalon. The frontal section (B) is located 3.5 mm rostral to the A–P zero point. The left half of the plate is a photograph of the brain section; the right half is a drawing that more clearly shows the structures outlined. Selected abbreviations of labeled structures that appear on the sagittal and frontal sections: am, medial nucleus of the amygdala; CC, corpus callosum; cho, optic chiasm; cf, column of the fornix; CI, inferior colliculus; FBM, medial forebrain bundle; H, hippocampus; lat, lateral hypothalamic nucleus; MED, medulla; MM, mammillary body; NC, neocortex; TH, thalamus; tro, optic tract; V, ventral nuclei of the thalamus. (Massopust, 1961, pp. 185, 198. From *Electrical Stimulation of the Brain: An Interdisciplinary Survey of Neurobehavioral Integrative Systems* edited by Daniel E. Sheer, © 1961, renewed 1989. By permission of the University of Texas Press.)

The Cathode Ray Oscilloscope:
Joseph Erlanger and Herbert Gasser

Prior to the 1920s, the technology of electrical recording from the brain was adequate for EEG recordings that essentially show the summated electrical activity of large populations of neurons. Since EEG waveforms range in frequency from 0.5 Hz to 50–60 Hz, the limitations of the early EEG machines, which could detect frequencies to only about 100 Hz, did not hinder research. A major advance in our knowledge of how the nervous system works at the cellular level awaited the development of technology for the recording of neural activity from single nerve cells. This required an instrument that could follow the voltage change of a single-cell action potential, a rapid transient lasting about one millisecond. The technological advance that permitted this occurred in the 1920s with the development of the vacuum tube for electronic amplification of voltage and for the acceleration and focusing of an electron beam onto the face of a cathode ray tube.

The vacuum tube voltage amplifier was a great improvement over earlier types of amplifier. The vacuum tube was also the basis for the development of the cathode ray oscilloscope. This is essentially a cathode ray vacuum tube with added circuits for voltage amplification and for sweeping the electron beam across the face of the tube from left to right at various speeds—and back again very rapidly. Early users of both these technologies were Joseph Erlanger (1875–1965) and Herbert Gasser (1888–1963) at Washington University in St. Louis, Missouri. Their published papers in 1922 and 1924 showed, for the first time, traces from the face of a cathode ray oscilloscope of compound action potentials recorded from the sciatic nerve of a frog and the tibial nerve of a dog (Gasser and Erlanger, 1922; Erlanger and Gasser, 1924). These 1922 and 1924 papers were pioneering publications in neurophysiology. Figure 2.5, adapted from an illustration in a book by Erlanger and Gasser (1937), shows the basic circuitry of an oscilloscope used in their early experiments.

The vacuum tube amplifier was demonstrated to be a stable and easy-to-use device for detecting in neural tissue very low level electrical signals of the sort first detected by Du Bois-Reymond. Amplification methods to detect the electrical nature of nerve activity had been available since the nineteenth century, but these techniques were relatively insensitive, unstable, and difficult to use. The vacuum tube amplifier made the electrical activity from nerves very accessible to detection. In addition, by using the cathode ray oscilloscope, Erlanger

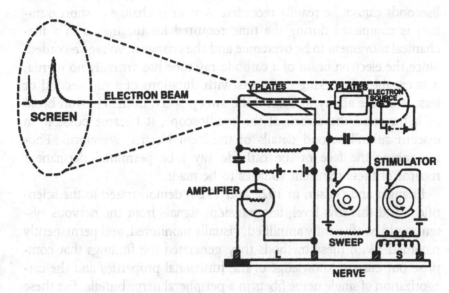

FIGURE 2.5. The cathode ray oscilloscope: a simplified diagram of the apparatus used by Erlanger and Gasser. The face of the oscilloscope is represented with a recording from a nerve displayed on it. The nerve preparation (e.g., an excised sciatic nerve from a frog) is in a recording chamber making electrical contact with two pairs of electrodes. An induction coil stimulator delivers an electrical shock via the stimulating electrodes (S) to initiate a compound action potential in the nerve. The neural response travels down the nerve and is recorded at the first of the two recording leads (L). The distal electrode is on the killed end of the nerve and serves as a reference point for the active recording lead. The input signal from the recording electrodes is connected via an amplifier to the Y deflection plates of the oscilloscope. The imposed neural voltage will deflect the electron beam up and down (in the Y dimension) on the face of the "scope." The sweep of the beam across the face of the oscilloscope had been triggered by the sweep generator that in turn had been activated by the stimulus signal. The first sharp deflection at the beginning of the oscilloscope trace is the stimulus artifact (so called because it is not a neural response) produced by the electrical shock to the nerve. The signal from the sweep generator is a voltage applied to the X deflection plates (so called because this sweeps the electron beam across the face of the oscilloscope from left to right, in the X dimension). (Erlanger and Gasser, 1937, p. 4. From *Electrical Signs of Nervous Activity* by J. Erlanger and H. Grasser. © 1937 University of Pennsylvania Press. Reprinted with permission.)

and Gasser were able to obtain accurate records of fast voltage changes in neural activity.

These scientists were the first to record and measure action potentials from peripheral nerves that consisted of fibers of different diameters and conduction velocities. This required more than amplification. The most sensitive early induction coil galvanometers used by Du Bois-Reymond, Caton, and Beck were just barely able to detect the presence of such fast transient signals and provided little information other than that the neural signal had occurred. Mechanical meter movements have so much inertia that voltage transients lasting only a few milliseconds cannot be readily recorded. A voltage change of short duration is completed during the time required for the inertia in a mechanical movement to be overcome and the change in voltage recorded. Since the electron beam of a cathode ray tube has virtually no inertia, it is capable of tracking transients with durations of a millisecond or less. With the ability to control the sweep speed of the electron beam as it traversed the screen of the oscilloscope, it became possible to discern submillisecond details of the neural signal waveform. Photographing the face of the cathode ray tube permitted permanent records of these transient voltages to be made.

Erlanger and Gasser, in 1922 and 1924, demonstrated to the scientific world that low level, fast transient signals from the nervous system could be efficiently amplified, visually monitored, and permanently recorded. With these methods they generated the findings that comprise our current knowledge of the functional properties and the categorization of single nerve fibers in a peripheral nerve bundle. For these important contributions, Erlanger and Gasser were awarded the Nobel Prize in Physiology or Medicine in 1944.

The cathode ray oscilloscope and sensitive electronic amplifiers were soon used for further advances in neurophysiology. Most important was the use of these instruments with very small-tipped microelectrodes, which served as probes for recording the electrical activity of single nerve cells. For an intracellular recording from a vertebrate neuron, the size of the microelectrode recording tip must not exceed one micrometer.*

It cannot be emphasized too strongly that many of the major advances in brain science could not have occurred without the technological advances that produced the electronic voltage amplifier, the cathode ray oscilloscope, and the stereotaxic instrument. This is certainly true for the scientific advances in sleep and waking research that are described in the following sections of the book.

---

*One micrometer (1 μm) is one thousandth of a millimeter (0.001 mm).

3

Early Research on Brain Mechanisms of Sleep and Waking

Encephalitis Lethargica: Constantin von Economo

During the final years of World War I and for a short time thereafter (1917–1919), a worldwide influenza epidemic was caused by one of the most deadly viruses ever known. The disease had a number of names, the best known being the Spanish flu. It has been estimated that between 25 and 40 million fatalities were due to the influenza epidemic—many more than the 9 million killed by the war. A variant of the disease was encephalitis lethargica, in which victims entered a coma that usually ended in death. Some of those who survived entered a rigid Parkinson-like state that was dramatically described by the neurologist Oliver Sacks, in his book *Awakenings* (1973), and strikingly portrayed in a Hollywood film of the same name.

Constantin von Economo (1876–1931) (Figure 3.1), a psychiatrist and neurologist who studied and worked in Vienna, saw a number of patients with the disease in 1917 and 1918 at the beginning and height of the epidemic. Von Economo was also a skilled neurohistologist, and over a number of years he analyzed the brains of those who had died of the disease, having failed to awaken from a sleep coma. He first described his findings in a series of papers beginning several years after the epidemic had peaked (von Economo, 1923). Brains were cut into thin sections, mounted on microscope slides and processed through various chemical solutions. One of the solutions contained a blue dye, the Nissl stain, which is taken up selectively by the cell bodies of neurons. When the Nissl-stained brain sections were studied under the microscope, von Economo discovered there was a profound loss of nerve cells in the posterior hypothalamus, located at the base of the diencephalon, and a region adjacent and caudal to it, the mesencephalic

FIGURE 3.1. Constantin von Economo working at the microscope. (Photograph kindly provided by the Institut für Geschichte der Medizin der Universität Wien and is reproduced with permission.)

reticular formation (von Economo, 1930). He also studied the brains of the small number of patients who exhibited essentially the opposite syndrome, a profound insomnia—an inability to sleep. The brains of these patients showed a cell loss in the anterior hypothalamus and the region adjacent and rostral to it, the preoptic basal forebrain area. Refer to Figure 1.2 to locate these regions at the base of the telencephalon and diencephalon.

These findings were the first substantial evidence pointing to specific brain areas responsible for sleep and waking. The posterior hypothalamus and mesencephalic reticular formation (MRF) appeared to be necessary for waking and the anterior hypothalamus and preoptic region necessary for sleep. To this day, von Economo's conclusions, with various refinements, are accepted as accurate.

Two Surgical Preparations: Frederick Bremer

Frederick Bremer (1892–1982) had a long scientific career in Brussels and made numerous contributions to neurobiology. In 1935 and 1936, this pioneering researcher (Figure 3.2) described two surgical preparations that have since been used to investigate many issues in neurobiology (Bremer, 1935, 1936). One of these preparations, first developed for use with cats, is the *encéphale isolé*. In this preparation the head and the brain are essentially isolated from the rest of the body. This is accomplished by a transection of the neural axis between the brain stem and spinal cord at the first or second cervical level, C1 or C2, (Figure 3.3B). The essential arterial and venous blood vessels that service the brain are, of course, left intact. The procedure is normally done under general gaseous anesthetic (ether was used at the time).

Figure 3.2. Photographic portrait of Frederick Bremer. (Reproduced with permission from E. D. Adrian, dedication to Professor Bremer. In: Progress in Brain Research, Vol. 1, Brain Mechanisms, G. Moruzzi, A. Fessard, and H. H. Jasper, eds. p. XIV. Amsterdam: Elsevier, 1963.)

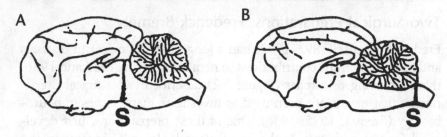

FIGURE 3.3. Representations of Bremer's *encéphale* and *cerveau isolé* preparations: (A) a sagittal section of a cat brain, showing a transection at the mesencephalic level to produce a *cerveau isolé* preparation, and (B) the transection between the medulla and spinal cord that produces the *encéphale isolé* preparation. (Representations adapted with permission from Bremer, 1936, p. 461.)

Immediately after the transection is made, the animal must be artificially respired because the central neural control to the muscles of the diaphragm that control breathing, as well as to all other muscles below the section, are observed. If room air (without anesthetic gas) is delivered to the lungs, the animal will soon regain consciousness. Before the animal regains consciousness, a local anesthetic is applied to wound edges and pressure points above the section to prevent painful input from arriving via cranial nerves. Having done all this carefully, and using the early technology of electroencephalography, Bremer demonstrated that the animal cycled through apparently normal sleep–waking states.

This preparation showed, not surprisingly, that the neural mechanisms responsible for sleep and waking did not require the spinal cord. Bremer's second surgical preparation, the *cerveau isolé* (Figure 3.3A), provided more interesting information. This is an isolated cerebrum produced by a transection of the neuraxis at the level of the rostral mesencephalon. The transection between the mesencephalon and diencephalon separates the forebrain from the brain stem. The forebrain (telencephalon and diencephalon) in front of the cut was intact, as was the brain stem and the rest of the nervous system behind the cut. Bremer observed that the forebrain of the *cerveau isolé* remained in a constant state of sleep. He pointed out that all sensory nerves except the ones for olfaction and vision are behind the section and do not have access to the thalamus and cerebral cortex. Figure 3.3 is from one of Bremer's original papers in 1936 that showed the location of transections (S in the drawings) that produced the *encéphale* and *cerveau* preparations.

Bremer interpreted these findings to mean that arousal or waking requires sensory stimulation and that sleep is produced and maintained by a sensory deafferentation of the cerebrum. Bremer's view implies that sleep is the baseline state of the brain. If there is a minimum of sensory stimulation or if an animal is deprived of sensory input, the brain will revert to its slow-wave sleep state. This is supported by the common observation that the conditions conducive to sleep are lying down in a quiet, dark room of normal temperature; essentially, the cerebrum's sensory input, including the proprioceptive feedback resulting from reduced muscle tonus in antigravity muscles, is then much reduced. This view of the neural basis of sleep has been called the passive deactivation theory of sleep. Another interpretation of Bremer's finding, not offered by Bremer at the time, is that areas of the brain necessary for a waking cerebrum are below the transection of the *cerveau isolé*. We will return to this issue in discussions of slow-wave sleep and waking.

In addition to Bremer's findings specific to the *cerveau* and *encéphale isolé* preparations, his approach served as an experimental model that was put to very valuable use by others. Subsequent workers severed the brain stem at levels between that of the *cerveau* and *encéphale* preparations to parcel out the brain stem regions critical for REM sleep. This research is described in Chapter 12, The Neural Control of REM Sleep.

An Interruption at Midcentury: World War II

By the 1930s, the work of von Economo and Bremer had provided tantalizing leads about the subcortical regions of the brain involved in sleep and waking. This area of research was centered in Europe. Also by the 1930s, electroencephalography and advances in electronics and in stereotaxic technique had developed to the point that these instruments and techniques were available for exploitation and ready to come together for use in brain research. However, in the 1930s conditions in Europe in everyday life as well as in science were severely disrupted, not only by the economic depression of the time, but by the rise of National Socialism in Germany. War broke out in Europe in 1939, bringing almost all non-military-related science to a halt in England and on the continent and, subsequently, throughout the world. If not for World War II, the neuroscience developments in sleep and waking research of the late 1940s and 1950s might well have occurred a decade earlier.

Part II
The Second
Half-Century: The
Benefits Are Reaped

Section 1
The Waking Brain

4

The Discovery of the Ascending Reticular Activating System

Shortly after war ended in 1945, publications appeared utilizing the available technologies that had been on hold since the 1930s. The first notable postwar research in sleep and waking came from a collaboration between an Italian and an American. In 1948 Giuseppe Moruzzi (1910–1986) at the University of Pisa in Italy, went to Northwestern University in Evanston, Illinois, to work for a year with Horace Magoun (1907–1991). The result of this collaboration was published in 1949 in the inaugural volume of the scientific journal, *Electroencephalography and Clinical Neurophysiology* (commonly known as "the *EEG Journal*") (Moruzzi and Magoun, 1949). In the study, entitled Brain Stem Reticular Formation and Activation of the EEG, cats were surgically prepared using a general anesthetic, α-chloralose, or the *encéphale isolé* preparation of Bremer. Both experimental procedures achieved the same effect, namely, a cat that displayed high voltage, low frequency (HVLF) EEG waves (the major sign of slow-wave sleep) that could be desynchronized into a low voltage, high frequency (LVHF) rhythm (the EEG sign of arousal). With the cat's head positioned in a stereotaxic instrument, subcortical brain areas could be accurately probed with electrodes for electrical stimulation.*

The original intent of the Moruzzi and Magoun study was to stimulate the superior cerebellar peduncle, the neural pathway from the cerebellum that courses through the rostral pons and mesencephalon

_____

*These preparations are referred to as "acute preparations" in the sense that all the data are collected in one session. At the end of the experimental session, the animal is sacrificed and the brain removed for histological analysis to determine the exact location of the electrode.

on its way to the nucleus ventralis lateralis, the motor system-related nucleus of the thalamus. The ventralis lateralis then relays the cerebellar influence to motor cortex. Moruzzi and Magoun intended to study the inhibitory effects of this stimulation upon motor cortex activity and motor cortex-induced movement. The fibers of the superior cerebellar peduncle in this region of the brain stem border upon and are interspersed with cells of the mesencephalic and pontine reticular formation. They were essentially stimulating the reticular formation. During the experiment, Moruzzi and Magoun monitored the EEG from the cerebral cortex. When the cat showed a synchronous EEG rhythm (indicating a sleeping brain), and electrical stimulation was delivered, Moruzzi and Magoun discovered that the cortical EEG immediately desynchronized into a pattern of low voltage fast activity. Of great interest was the fact that this activation pattern was not limited to frontal/motor cortex, where an effect from stimulating a cerebellar–thalamic–motor cortex circuit was expected; instead, it was seen over the entire cortical surface—indicating an aroused cerebral cortex.

Moruzzi and Magoun determined the stimulation voltage that would evoke a threshold arousal response and noted that the EEG desynchronization lasted only as long as the stimulation period (3–5 seconds).

FIGURE 4.1. Photograph of Moruzzi (left) and Magoun taken in 1958 in Warsaw, Poland, on a return trip from Moscow where they had attended a colloquium. (Reproduced with permission from H. W. Magoun's personal and professional papers donated to the UCLA Neuroscience History Archives.)

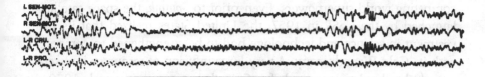

FIGURE 4.2. Cortical activation produced by stimulation of the reticular formation in the cat. This figure, from Moruzzi and Magoun's landmark paper of 1949, shows four channels of EEG recordings from an *encéphale isolé* preparation with a small amount of anesthetic administered to facilitate the appearance of EEG synchronization. The black line at the bottom marks the onset and duration of electrical stimulation to the reticular formation. It can readily be seen that the stimulation desynchronized the EEG to produce a low voltage fast activity pattern characteristic of an activated cortex. Shortly after the offset of stimulation, the EEG returned to slow-wave activity. (Moruzzi and Magoun, 1949, p. 456. Adapted from Moruzzi G, Magoun HW (1949). Brain stem reticular formation and activation of the EEG. Electroencephalogr Clin Neurophysiol 1:455–473. © 1949 with permission from Elsevier Science.)

As the stimulus strength was increased, the arousal response became progressively longer. They were able to titrate the amount (duration) of arousal as a function of stimulation intensity. By this procedure, Moruzzi and Magoun showed that the reticular formation exerted a precise and tight control over arousal. Figure 4.1 shows the two collaborators in 1958 during a visit to the Nencki Institute in Warsaw, Poland. Figure 4.2, from their original paper in 1949, illustrates the EEG activation produced by reticular formation stimulation.

Subsequent studies using unanesthetized, behaving animals with electrodes implanted for long-term stimulation and recording confirmed that the investgators had indeed discovered a brain region that, with gentle pulsing, would awaken a sleeping animal.* As Magoun states in his book *The Waking Brain*, "Evocation of this generalized and self-maintaining electrocortical alteration has since become commonplace, but it is still possible to recall the arousal evoked in the investigators by its initial display!" (Magoun, 1963, p. 27). The reticular formation and neural circuits associated with the arousal response were soon to become known as the ascending reticular activating system (ARAS), probably the most important heuristic concept in the entire sleep–waking field.

This publication in 1949 had immense influence. The end of the war saw the emergence of the United States and the Soviet Union as two

---

*These animals, in contrast to acute preparations, are referred to as chronic preparations. The electrode implant surgery is done under general anesthesia and the animal is permitted to recover. The chronic preparation is then used in many experimental sessions that may occur over periods of days, weeks, or months.

competing world superpowers. Even prior to *Sputnik*, the space satellite launched by the USSR in 1957, the United States realized that its investment in science and technology had to be strengthened to compete with its Cold War adversary. A very large infusion of resources into science and technology occurred after *Sputnik*, but even in the early 1950s, before *Sputnik*, under conditions of postwar economic prosperity in America, money was available to nourish the research of scientists in all fields. At this opportune time, the study by Moruzzi and Magoun had the effect of coalescing the interests of scientists from disparate fields to pursue interdisciplinary research related to the mechanisms of arousal.

Concepts like arousal and attention have always been of interest to psychologists. Those with backgrounds in brain physiology and anatomy realized that they now had the opportunity to explore the biological bases of these phenomena. Those with little background in biology were encouraged to learn more. Similarly, neurophysiologists and neuroanatomists were stimulated to become familiar with the study of behavior, the main consequence of neural activity in the awake brain. It is important to note that a major function of waking is to provide a brain state that supports behavior. Interactions between psychologists, physiologists, and anatomists were stimulated by the work of Moruzzi and Magoun as well as by developments in areas such as limbic system physiology. Interactions and collaborations like these eventually developed into the thriving interdisciplinary field we now call neuroscience.

Moruzzi returned to Pisa soon after the groundbreaking paper was published, and around the same time Magoun left Northwestern University to establish the Department of Anatomy at the University of California at Los Angeles (UCLA). Based on their individual research accomplishments and collaborative pioneering work—in addition to each man's personal charisma—they established research centers that attracted scientists from all over the world. Certainly not all the work at these facilities was devoted to the brain stem control of sleep and waking, but much of it was. Moruzzi's institute at Pisa concentrated more on sleep, and this work will be described in a later section; the UCLA group initially pursued the waking part of the story. One of the first workers to collaborate with Magoun in California in the early 1950s was a neurosurgeon, John French (1911–1989). He soon became very active with Magoun in raising funds to establish a research institute and training program at UCLA—and he became its first director. In 1961 a 10-story building dedicated to brain research, the Brain Research Institute (BRI), was completed on the UCLA campus. This building was soon filled with brain scientists, many of whom worked on the ARAS.

In addition to the cadre of international scientists recruited to its staff, the BRI established a postdoctoral training program funded by the National Institute of Mental Health (NIMH). In 1960 I was selected as a BRI postdoctoral trainee in the Department of Physiology. A number of the trainees took the opportunity to attend courses at the medical school. In a neuroanatomy brain dissection lab, Professor Magoun was the laboratory instructor for the postdoctoral trainees and graduate students. There was general excitement when Magoun made a midsagittal cut through an intact human brain and cupped a complete half-brain in one hand, with its medial surface exposed toward the students. With a probe in the other hand, Magoun pointed out the region of the reticular formation that he and Moruzzi had stimulated to produce EEG arousal. To use a phrase similar to one used by Magoun when he described his and Moruzzi's finding, I recall the arousal evoked in us by the display of the reticular formation by the codiscoverer of one of its important functions!

Donald Lindsley, another early collaborator of Magoun's, was a physiological psychologist at Northwestern University while Magoun was there and joined Magoun at UCLA in 1951. While still at Northwestern, Lindsley, Magoun, and coworkers had completed two important experiments on the ARAS. The first was published as a companion paper to the Moruzzi and Magoun article in the first volume of "the EEG Journal"; the second paper appeared in the second volume of that journal (Lindsley et al., 1949, 1950). Both papers showed the effects of lesioning the rostral region of the reticular formation. The lesioned area included the region stimulated by Moruzzi and Magoun and the area in which von Economo had found cell loss in patients who had suffered from encephalitis lethargica. Lindsley, using experimental animals, controlled the lesion size and location that was clinically produced in humans by an immune reaction to a viral infection that affected a restricted population of nerve cells. (It is interesting to note that years after the flu epidemic it was discovered that the cause of the clinical symptoms was cell death caused not by the virus itself, but by the body's immune response to cells that harbored the virus.)

Lindsley and his coworkers (1949) used an unusual approach to destroy the rostral reticular formation. To avoid damaging the cortex with the multiple penetrations required to lesion the entire region with a conventional dorsal approach, they left the overlying skull intact and instead exposed the cerebellum. The lesion electrode penetrated the cerebellum and was advanced at almost a horizontal angle until the tip reached its target. At this location, the electrode delivered a precise amount of current to destroy brain tissue around it. The electrode was repositioned so that the adjacent area of brain was lesioned. The process was repeated until the entire rostral reticular formation was destroyed. Using this approach through the cerebellum, the cerebral cortex, from which subsequent EEG recordings were taken, was left intact and uninjured. Figure 4.3, reproduced from the 1949 paper, shows the extent of the reticular formation lesions.

The first of the two papers reported that acutely prepared cats with electrolytic lesions of the reticular formation (Figure 4.3C, D) showed cortical EEG slow waves continuously for the duration of time they were monitored (Figure 4.3G). This finding was not surprising. It was consistent with von Economo's finding of sleep coma with clinical damage and with the discovery by Moruzzi and Magoun that *activation* of this tissue caused arousal.

To make certain that the effect on the reticular formation of a large lesion was specific to this population of destroyed cells and could not be attributed merely to a relatively large area of brain destroyed there or elsewhere, another manipulation was added. Control lesions of a similar size were made dorsal to the reticular formation and on each side of the brain stem lateral to the reticular core. The reticular formation ventral and medial to these control lesions was spared. Figure 4.3 shows the extent of the dorsal (A) and lateral (B) lesions. The control-lesioned cats (with the reticular formation intact) showed sustained periods of cortical EEG arousal (Figure 4.3E, F). Lindsley et al. concluded that the loss of arousal and the induction of a slow-wave sleep coma were specific to destruction of the mesencephalic reticular core.

The second of the two papers by Lindsley et al. (1950) was more interesting in that the investigators used chronically prepared cats, which allowed the observation of behavior. Reticular formation and control lesions, comparable to those made in the first paper, were produced under surgical anesthesia. The animals were then carefully nursed during recovery periods lasting as long as two months. In keeping with the human data from encephalitis lethargica patients, the cats with destruction of the reticular core were in a comatose state. When the control-lesioned animals recovered from the anesthetic and surgery, they showed normal sleep and waking behavior. This control experiment demonstrated that the behavioral as well as the EEG effects of reticular formation lesions were, in fact, specific to the reticular formation. However, additional observations on these cats proved to be even more interesting and important.

The expectation at the time was that the control-lesioned animals would not be responsive to touch and auditory stimuli, since the lateral lesions to the brain stem (Figure 4.3B) had destroyed the ascend-

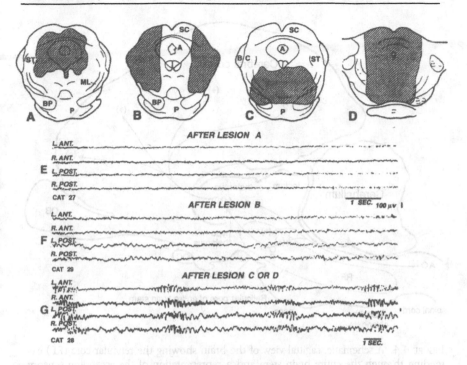

FIGURE 4.3. Lesions of the reticular formation and areas lateral and dorsal to the reticular core, and subsequent EEG tracings following these lesions. (A–D) Frontal sections through the regions of the midbrain that were lesioned in cats after initial transections were made between the medulla and spinal cord (*encéphale isolé* preparations). These cats, prior to the brain stem lesions, showed periods of EEG cortical activation interspersed with episodes of EEG synchrony. (C, D) Brain sections from two cats with large lesions that destroyed the reticular formation. The EEG traces in (G), recorded after the lesion illustrated in (C), show persistent recurring spindle bursts, a form of EEG synchrony seen in slow-wave sleep. The lesion shown in (D) was followed by similar EEG spindle activity. The control lesions of comparable size but dorsal (A) and lateral (B) to the reticular core resulted in EEG records with sustained periods of EEG activation (E, F). The EEG tracings in the original 1949 journal article are degraded; however, the main points of the presence of low voltage desynchronization in traces (E) and (F) and of high voltage synchrony in the traces in (G) are readily seen. (Lindsley et al., 1949, p. 477. Reproduced from Lindsley DB, Bowden J, Magoun HW (1949) Effect upon the EEG of acute injury to the brain stem activating system. Electroencephalogr Clin Neurophysiol 1:475–486. © 1949 with permission from Elsevier Science.)

ing medial and lateral lemniscal pathways that, at the mesencephalic level, are located laterally. These pathways carry tactile and auditory signals from their peripheral origins to their specific sensory nuclei of the thalamus, the ventral posterior nuclei for touch, and the medial geniculate nucleus for audition. However, despite destruction of these large direct fiber tracts that carry sensory signals to the thalamus, which in turn relays the sensory messages to the cortex, the cats could be

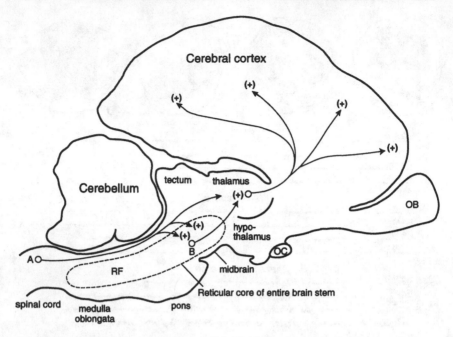

FIGURE 4.4. A schematic, sagittal view of the brain showing the reticular core (RF) extending through the entire brain stem and a representation of the ascending sensory pathways (A) to the thalamus. Also represented are sensory collaterals that excite cells of the reticular formation, and the reticulothalamic (B) and thalamocortical arousal pathway. Abbreviations: OB, olfactory bulb; OC, optic chiasm; RF, reticular formation.

awakened from sleep by auditory or tactile stimuli! How could a sensory message cause cortical arousal without the signal reaching the thalamus and then the cortex? Subsequent research provided an explanation for the mystery.

Lindsley et al. in 1950 had guessed at the answer. They speculated that as these direct lemniscal sensory pathways ascend the lateral brain stem to the thalamus, collateral branches from the primary axons veer off medially to terminate in the brain stem reticular core. The lateral lesions prevent the primary axons from reaching the thalamus but the collaterals, in contrast, do reach the reticular formation. Thus, sensory inputs could reach and activate the reticular formation to cause arousal. Anatomical and electrophysiological research in the mid-1950s confirmed the accuracy of this speculation and revealed important physiological properties about the ARAS. Figure 4.4 is a schematic representation of the brain showing the sensory collaterals activating the reticular formation and the reticular–thalamic–cortical pathway that causes arousal. Information will be added to this basic figure as we learn more about arousal and sleep influences and where they exert their effects.

At the time of Lindsley's work, researchers were beginning to make recordings from cells of the mesencephalic reticular formation during the presentation of sensory stimuli. Two of these early studies were done in Moruzzi's lab in Pisa, by visiting scientists working with Moruzzi and his colleague Mollica. Arnold and Madge Scheibel were a husband-and-wife team who joined Magoun at UCLA after their stay in Italy, and Philip Bradley was a neuropharmacologist from the University of Birmingham, England. Both experiments used the then-recent technique of recording action potentials from single nerve cells without having to penetrate the cell with an intracellular microelectrode (Scheibel et al., 1955; Bradley and Mollica, 1958). In experiments of these types, a microelectrode is advanced into the brain until the area of interest is reached. A micromanipulator on the probe drive is then used to advance the electrode a few micrometers at a time until the very fine electrode tip is close enough to a single nerve cell to "see" its action potentials in relative isolation from the action potentials of neighboring cells. Figure 4.5 shows an oscilloscope trace of a typical extracellular single unit recording from a cell in the mesencephalic reticular formation.

In the studies conducted at Pisa, one of the interesting observations was that a number of cells of the reticular formation responded to more than one sensory modality. The researchers showed, electrophysiologically, a convergence of sensory inputs upon individual reticular neurons. From the firing pattern of the neuron, it was not possible to discern the nature of the sensory stimulus that "fired" the cell, thus suggesting that information about the specific sense modality that activated the neuron is, apparently, lost. This last point explains certain observations made of animals with lemniscal lesions. When awakened

100 ms ⊢⊣

FIGURE 4.5. Photograph of an oscilloscope trace showing extracellular spike action potentials recorded from a single cell of the mesencephalic reticular formation of a cat. The firing rate of this cell is about 40 spikes per second. Detailed characteristics of individual action potentials, such as waveform and duration, can be seen by increasing the sweep speed and amplification of the oscilloscope trace. (Adapted with permission from J. Siegel and Lineberry, 1968, p. 453.)

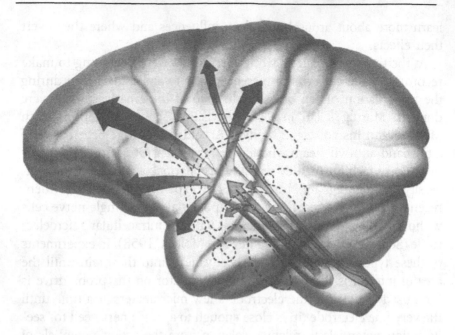

FIGURE 4.6. Magoun's 1954 representation of the ascending reticular activating system displayed on a lateral view of the monkey brain. This figure shows the reticular formation activated by collateral branches of sensory pathways going to the specific sensory nuclei of the thalamus. The reticular formation, in turn, activates thalamic nuclei that project diffusely to the entire cortex (the large, dark arrows) to produce cortical arousal. (Reproduced with permission from Magoun, 1954, p. 13.)

by a sensory stimulus, for example, a touch to a part of the body, the cats do so in a manner different from that of normal, nonlesioned animals. Intact animals, when awakened, orient to where on the body the tactile stimulus occurred. Sensory nuclei of the thalamus are organized to encode detailed information about the location of sensory stimuli. Animals with lemniscal pathway lesions also awaken, but they orient in a manner indicating that the information about the specific nature of the sensory stimulus is lost. From this difference it may be concluded that awakening to a sensory stimulus is accomplished by an activation of the reticular formation that, in turn, produces a nonspecific, generalized arousal of the cortex. The cortex, now in an aroused state, is capable of processing and reacting to the detailed sensory information that arrives over the classical specific sensory pathways. That is, there are two parallel sensory input pathways that serve different functions with respect to the cerebral cortex.

At the time of the work by Moruzzi, Magoun, and Lindsley, it was not known that the rostral projections from the reticular formation do not course directly to the cerebral cortex. Since Moruzzi and Magoun had shown that incremental cortical arousal effects were tightly regulated by incremental changes of reticular stimulation, some researchers assumed that there was a direct cortical projection. However, Moruzzi and Magoun, as early as 1949, had suggested the possibility that the generalized cortical arousal may be "mediated, in part, at least, by the diffuse thalamic projection system" (Moruzzi and Magoun, 1949, p. 468). This possibility was graphically represented in a paper by Magoun (1954) in which a schematic view of the brain showed a projection from the reticular formation to nuclei of the thalamus that, in turn, projects diffusely to the cortex. This drawing is shown in Figure 4.6.

5

The Neural Pathways
That Produce Arousal

As anatomical and physiological evidence accumulated in the wake of the publication by Moruzzi and Magoun (1949), it was eventually found that cortical arousal produced by the ARAS originates in certain nuclei of the rostral reticular formation and is mediated by two ascending pathways, one dorsal and one ventral (Jones, 2000). A major source of the arousal system resides in cells of two nuclei, the pedunculopontine tegmental nucleus (PPT) and the laterodorsal tegmental nucleus (LDT). The cells of these nuclei are located at the pontine–mesencephalic junction, where they essentially surround and intermingle with fibers of the superior cerebellar peduncle. Recall that in the experiment of 1949, Moruzzi and Magoun used the superior cerebellar peduncle as the target of the stimulating electrode. This peduncle is also called the brachium conjunctivum, and the nuclei that are adjacent are often referred to as the peribrachial nuclei.* It has been shown that these two nuclei contain acetylcholine-synthesizing neurons; we know, therefore, that the major cells of origin of the ascending arousal system are cholinergic (Jones and Beaudet, 1987; Jones and Webster, 1988). Figure 5.1 shows a frontal section through the catbrain and the location of the cholinergic cells of the peribrachial nuclei.

The dorsal pathway from the peribrachial nuclei was the one anticipated by Moruzzi and Magoun in 1949. This involves a projection to the intralaminar and midline nuclei of the thalamus, which in turn projects broadly to the cortex (Starzl et al., 1951; Nauta and Kuypers,

---

*A cell population just caudal to the peribrachial nuclei and also adjacent to the brachium conjunctivum is called the *para*brachial nucleus (Steriade and McCarley, 1990, p. 69).

47

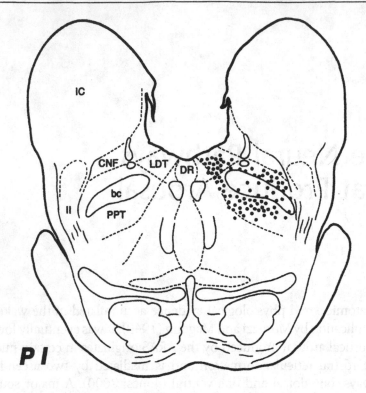

FIGURE 5.1. A frontal section through the cat brain at the pontomesencephalic junction showing the location of the peribrachial nuclei. Small solid circles on the right represent cells that stained positive for choline acetyltransferase, an enzyme that is present in cholinergic neurons and serves as a marker for such neurons. Abbreviations: bc, brachium conjunctivum; CNF, cuneiform nucleus; DR, dorsal raphé nucleus; IC, inferior colliculus; LDT, laterodorsal tegmental nucleus; ll, lateral lemniscus; PPT, peduculopontine tegmental nucleus. (Jones and Beaudet, 1987, p. 22. Adapted from Jones BE, Beaudet A (1987) Distribution of acetylcholine and catecholamine neurons in the cat brain stem studied by choline acetyltransferase and tyrosine hydroxylase immunohistochemistry. J Comp Neurol 261:15–32. © 1987 by permission of Wiley-Liss, Inc., a subsidiary of John Wiley & Sons, Inc.)

1958; Saper and Loewy, 1980; Macchi and Bentivoglio, 1986). There is evidence that the thalamic projection to the cortex is mediated by glutamate, an excitatory amino acid neurotransmitter. As early as 1961, Herbert Jasper, at McGill University in Montreal, described a ventral projection from the reticular formation in addition to the dorsal pathway through the thalamus. Jasper stated that the ventral limb bypassed the thalamus and projected to the hypothalamus and continued more rostrally into the septal region of the basal telencephalon (Jasper, 1961). More recently, Barbara Jones and her colleagues, also at McGill University, using modern anatomical tracing techniques, detailed this pathway (Jones and Yang, 1985; Jones, 1990, 1993).

The major nuclei and trajectories of the dorsal and ventral components of the ascending arousal system that originate in the rostral reticular formation (pontomesencephalic junction) are shown in Figure 5.2.
At the diencephalic level, prior to reaching the thalamus, components
of the fiber tract from the peribrachial nuclei of the reticular formation
veer off from the dorsal bundle and course ventrally to join a major
fiber tract, the medial forebrain bundle, that courses through the ventral region of the forebrain. Axons in this bundle from the reticular formation have a number of forebrain targets, including nuclei of the hypothalamus in the diencephalon and nuclei of the basal forebrain in
the telencephalon.

In this and subsequent chapters, the effects of neurotransmitters will
be discussed. We should understand certain characteristics of neurotransmitters, hormones and receptors. Neurotransmitters are synthe-

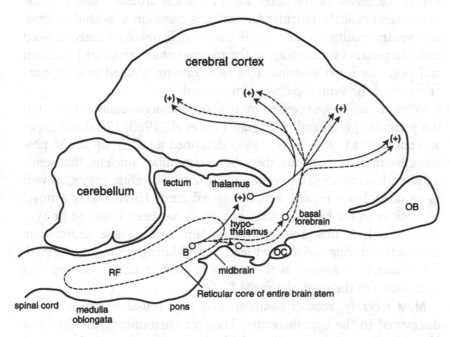

FIGURE 5.2. The major nuclei and trajectories of the dorsal and ventral components of
the ascending reticular activating system that originate in cells of the pontomesencephalic brain stem (B). The dorsal glutamatergic pathway from the intralaminar and
midline nuclei of the thalamus to the cerebral cortex is shown with solid lines. The
divergent, branching lines represent the diffuse nature of this thalamocortical projection system. The ventral pathway, represented by dashed lines, includes a hypothalamic histaminergic component and a basal forebrain cholinergic component, both of
which also project broadly to the cerebral cortex. Abbreviations: OB, olfactory bulb;
OC, optic chiasm; RF, reticular formation. (Adapted with permission from Jones, 2000,
p. 138.)

sized by nerve cells and have their primary controlling effects or modulatory influences upon postsynaptic neurons. Hormones that affect
the nervous system are often referred to as neurohormones. Transmitters and hormones exert their effects on neurons by binding to receptor sites located on the nerve cell membrane. The particular type of receptor to which the chemical (ligand) binds determines the effect it
will have. A single neurotransmitter can have opposite effects (excitatory or inhibitory) by binding to different receptor types. Notable ex-
ceptions are certain transmitters that always have one type of effect—
either excitatory or inhibitory—on their postsynaptic targets. For
example, glutamate is always an excitatory neurotransmitter and
γ-aminobutyric acid (GABA) is an inhibitory transmitter.

The influential work of Moruzzi and Magoun, led to a most thor-
ough study of the reticular formation and the dorsal pathway to the
cortex through the diffuse projection nuclei of the thalamus. This system is the classic neural substrate for cortical arousal. Note that the
pontomesencephalic reticular formation is common to both the dorsal
and ventral pathways, and recall that von Economo observed sleep
coma in patients with lesions of the mesencephalic reticular formation
and posterior hypothalamus. This early finding pointed to an important role of the ventral pathway for arousal.

Work in the 1980s revealed more about the organization of the ventral pathway. Jouvet and colleagues (Lin et al., 1988, 1996) and Saper
and colleagues (1985, 1987, 1997) described a cell group in the posterior lateral hypothalamus, the tuberomammillary nucleus, that sends
a diffuse histaminergic projection to the entire cerebral cortex, as well
as to other brain regions and the spinal cord. Histamine is a novel
transmitter in the brain in that, thus far, it has been found to be synthesized only in this group of hypothalamic cells. These neurons are
most active during waking and least active during sleep. As expected,
the release of histamine at the cortex is greatest during waking and
contributes to the cortical arousal.*

Most recently, another component of the arousal system has been
discovered in the hypothalamus. The area surrounding the fornix (a
fiber bundle that courses through the hypothalamus) called the peri-
fornical area, and regions lateral to it in the hypothalamus, contain cells

---

*The drowsiness caused by antihistamine allergy medication appears to be due to its
widespread antagonistic action on histamine receptors, including those that are normally activated by the hypothalamic–cortical projection. This reduces the effectiveness
of the histaminergic component of the cortical arousal system.

that synthesize a newly discovered peptide neurotransmitter. This neuropeptide has been given two names, orexin and hypocretin (Kilduff and Peyron, 2000). (A more detailed description of this transmitter is in a section on narcolepsy in Chapter 14.) Hypocretin cells project to and have an excitatory effect on many arousal-producing areas of the brain as well as directly to the cerebral cortex (Peyron et al., 1998; Moore et al., 2001). Recent evidence indicates that hypocretin neurons are more widely distributed in the hypothalamus than previously thought (Hopkins et al., 2001) and are intermingled with the histaminergic neurons of the hypothalamus. Moreover, the tuberomammillary cells express hypocretin receptors and are excited by hypocretin—and hypocretin cells are synapsed upon by histaminergic neurons (Eriksson et al., 2001). It appears that the posterior hypothalamus houses two neuronal populations that cooperate in regulating arousal level.

The most rostral component of the ventral arousal system, prior to its termination in the cerebral cortex, is the basal forebrain in the telencephalon. Under the influence of the reticular input, nuclei of the basal forebrain become active. These nuclei (the diagonal band of Broca, the nucleus basalis, and the substantia innominata) contain acetylcholine-synthesizing neurons commingled with GABA-synthesizing neurons. Both types of neuron project to the cerebral cortex. Of the two, the cholinergic contingent provides the greater input to the cortex. Evidence has accumulated that the cholinergic influence facilitates the responsiveness of cortical neurons to sensory stimuli; that is, they are involved in cortical arousal.

We see, then, that in addition to the classical reticular–thalamic–cortical dorsal circuit for arousal, there is a ventral arousal circuit that has two components. There is a reticular–hypothalamic–cortical pathway and a reticular–basal forebrain–cortical pathway. It should also be noted that components of both the dorsal and ventral pathways have been found to be involved in sleep as well as in arousal. This is discussed in the chapters on sleep (Chapters 6–12).

In addition to the reticular formation–cortical influences that are mediated by subcortical nuclei of the dorsal and ventral pathways and their transmitters glutamate, acetylcholine, and histamine, there are other brain stem nuclei that project to the cortex. The terminals of these nuclei provide serotonergic, noradrenergic, and dopaminergic influences that also regulate waking and sleep (Saper, 1987).

The findings that parts of the thalamus, hypothalamus, and the basal forebrain participate in the function of cortical arousal helps us un-

derstand a body of data that, at an earlier time, was difficult to interpret. I refer to findings of recovery of arousal function after damage to
the reticular activating system. The early work of Bremer on the *cerveau
isolé*, in which a transection between the forebrain and brain stem left
the forebrain in a state of slow-wave sleep, suggested that a region behind the section was necessary for the forebrain to experience arousal.
Subsequently, the research of Moruzzi, Magoun, and Lindsley pointed
to the rostral reticular formation as that region. However, the brain
never ceases to inform us about its plasticity.

It has been demonstrated in a number of ways that there is some recovery of function under certain circumstances. Adametz (1959) reported that if the reticular formation is lesioned as extensively as was
done by Lindsley, but in stages over several days, the cats did not lose
arousal completely. Furthermore, *cerveau isolé* preparations that were
carefully nursed and kept viable for a period of three weeks eventually
showed some small recovery of arousal function (Villablanca, 1965).
The interpretation of these data has been that the reticular formation
is certainly an important region governing arousal, but it may not be
the only structure to subserve arousal. We now know that certain nuclei of the hypothalamus, thalamus, and basal forebrain, all structures
rostral to the section of the *cerveau isolé*, are components of the ascending arousal system. These forebrain components of the arousal system may have some capability to sustain arousal when the major brain
stem component has been destroyed.

These and similar findings in other areas have prompted workers
in brain physiology to shun the notion of brain centers controlling behavioral functions. The more appropriate conceptualization is that behavioral functions are based on systems of nuclei and interconnecting
pathways that comprise anatomically distributed neural circuits. This
view has been articulated in a paper by Jouvet in which the first two
sentences of the abstract state: "The hypothesis of a unique waking system located in the mesencephalic reticular formation is no longer ten-
able. Cortical and behavioral waking depends upon a complicated network of different systems which interact with each other with different
neurotransmitters" (Jouvet, 1996, p. 762).

Part II
The Second
Half-Century: The
Benefits Are Reaped

Section 2
The Sleeping Brain:
Slow-Wave Sleep

6

Forebrain and Hindbrain Inhibition of the Reticular Activating System

Arousal may be seen as having two components: ascending and descending. Ascending activation engages cortical processes for conscious, cognitive (psychological) arousal and the subsequent behavioral adjustments appropriate to the conditions that initiated the arousal. The descending component may be described as physiological activation in which the somatic motor system and the sympathetic component of the autonomic nervous system produce a complex of changes to support the increased activity that normally occurs during waking. It was recognized in the 1950s that both the ascending (psychological) and descending (physiological) components of arousal had elements of positive feedback to the brain stem arousal system. A number of workers in the UCLA group, as well as others, demonstrated that stimulation of certain cortical sites produced cortical arousal (Segundo et al., 1955a,b; Adey et al., 1957; Kaada and Johannessen, 1960). This effect was shown to be mediated by a corticoreticular influence that activated the ARAS; that is, the cortex does not possess the ability to arouse itself without the mediation of the ARAS. An everyday understanding of this corticoreticular influence would be the increased arousal produced by the daydreaming of stimulating thoughts.

Descending, physiological arousal involves a number of changes produced by the autonomic and endocrine systems. These include increased heart rate, blood pressure, and respiration, as well as the release of adrenal hormones. In addition, descending arousal produces increased tonus in skeletal muscle. Some of these peripheral autonomic, endocrine, and somatic motor changes generate positive feedback to the brain that result in greater descending activation—and so on. Figure 6.1 shows the schematic presented in Figure 4.4 but with the ad-

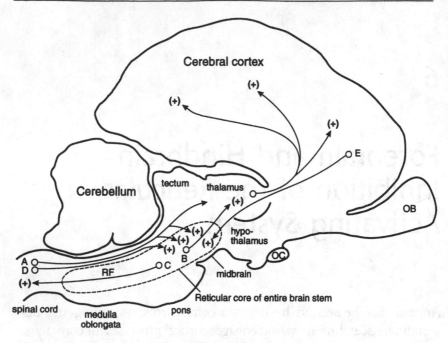

FIGURE 6.1. This figure duplicates Figure 4.4 with the addition of descending activation (C) and positive feedback to the reticular activating system from an aroused body (D) and cortex (E). Abbreviations: OB, olfactory bulb; OC, optic chiasm; RF, reticular formation.

dition of the descending component of arousal and the positive feedback on the arousal system from both an activated cortex and activated body.

The identification of positive feedback loops and their potentially accelerating, runaway responses requires that there be inhibitory or dampening influences to counter and control these effects. Inhibitory influences must exist, since the ultimate state of cortical activation, a generalized seizure, is not the inevitable outcome of awakening in the morning. The recognition and the study of inhibitory influences upon arousal occurred in the mid-1950s and ushered in an era of sleep research that led to our current level of understanding about the neural mechanisms that produce sleep. Pioneers in this work included Paul Dell, Marguerite Bonvallet, and their colleagues in Paris, as well as Giuseppe Moruzzi and his colleagues in Pisa.

Between 1954 and 1960, Dell, Bonvallet, and their colleagues published a remarkable series of papers that is summarized by Dell, Bonvallet, and Hugelin in a chapter of a 1961 publication on the nature of sleep (Dell et al., 1961). Dell points out that "it was Bremer (1935) who introduced the concept of sleep as a process of 'deafferentation' following his observations on the *cerveau isolé* preparation. Those who hold this point of view regard sleep as a process of *passive reticular deactivation*" (Dell et al., 1961, p. 86). There is no question, as pointed out earlier, that the removal of sensory input is conducive to sleep. Lying down in a dark, quiet, comfortable room of normal temperature is the condition most of us look for to fall asleep. This is a passive deactivation. However, passive deactivation, as Dell was to demonstrate, is not the whole story. Dell provided evidence that the ARAS receives *active* deactivating influences from a number of brain regions. Early work showed two major sources of active reticular deactivation to be at extreme opposite poles of the brain, the cerebral cortex and the lower (bulbar) brain stem.

As already discussed, the cortex, activated by reticular influences, can exert excitatory influences on the reticular formation that, in turn, increases the cortical arousal. In a series of papers, Hugelin and Bonvallet (1957a–c, 1958), presented evidence that the cortex also exerts descending *inhibitory* influences to the rostral reticular formation to counteract and regulate its facilitory actions. In these experiments Hugelin and Bonvallet utilized the monosynaptic masseteric stretch reflex, a jaw closure reflex, to demonstrate this inhibition. The sensory and motor branches of the trigeminal nerve control this reflex. The masseteric reflex was elicited by an electric shock to the appropriate nucleus of the trigeminal system and a baseline level of the amplitude of the reflex was established. Upon electrical stimulation to the reticular formation that produced increased arousal, a marked facilitation of the reflex was observed. However, if the reticular stimulation was continued and maintained at a constant level, the amplitude of the jaw reflex rapidly decreased to the prereticular stimulation level within a few successive trials. When the cerebral cortex was removed by lesion or inactivated temporarily by cooling, the reticular stimulation had the same initial facilitory effect; however, the reflex strength did not decrease but instead was maintained throughout the duration of the reticular stimulation.

This work from Dell's lab demonstrated that the intact cerebral cortex contains a mechanism capable of regulating its level of arousal by exerting, in this case, an inhibitory influence upon the ARAS. We recall that the cortex also has the ability to increase its arousal level by exerting an excitatory corticoreticular influence. Thus, it appears that the cortex can regulate its own level of arousal appropriate to the circumstances. For example, if the cortex processes incoming information as being irrelevant, boring, or meaningless, it would exert an inhibitory influence on the reticular formation to decrease its ascending activation and the individual might doze off—as can be attested to by many a student attending a boring and droning lecture. Nodding off is more probable under those circumstances than if the individual were in the same lecture hall without the lecture. The droning lecture may be considered comparable to the presentation of rhythmic meaningless sounds, like those of a beating metronome. Stimuli of that sort have a soporific (synchronogenic/sleep-inducing) effect. In contrast, if the student can effectively filter out the lecture and turn his or her mind to fantasizing about an event of interest, like an upcoming date, the cortical processing of this internally generated information would selectively engage the excitatory cortical influence to the reticular formation and thus produce an increased arousal level.

We turn now to influences from the lower brain stem that modulate arousal. A number of workers have shown that the bulbar brain stem can exert both descending and ascending inhibitory effects. Dell et al. (1961) demonstrated a tonic descending inhibitory effect on somatic motor responses by measuring the strength of a spinal reflex in a high spinal preparation (the *encéphale isolé,* with the neuraxis severed between the spinal cord and medulla) compared with a transection *rostral* to the medulla. Without the influence of the medulla of the lower brain stem, the spinal reflex was quite vigorous; with the medulla connected to the spinal cord, the spinal reflex was of considerably less magnitude. Interestingly, when the spinal preparation was compared with an intact nervous system in which the lower brain stem *and the rest of the brain in front of it* were connected to the spinal cord, the inhibitory influence of the bulbar brain stem was difficult to detect. This result has been attributed to inhibitory influences originating in the cortex. Prefrontal and orbitofrontal cortical areas were found to exert an inhibitory influence upon the inhibitory region of the lower brain stem (Dell et al., 1961). Conceptually, this is somewhat tricky, but it does illustrate a not uncommon device in the nervous system, that of disinhibition, or inhibition of inhibition. The end result is a neutralization of an inhibitory influence, which may appear as an excitation. Disinhibition occurs when it is inappropriate for an inhibitory process to operate or when the strength of an inhibitory effect needs to be decreased.

Identification of the area within the lower brain stem responsible for an ascending inhibitory influence was accomplished by using focal stimulation within the medulla. Pioneering work on this was done at the Institute of Physiology at Pisa. In a seminal paper presented at the

"Nature of Sleep" symposium in 1961, Moruzzi described research with coworkers Magnes and Pompeiano (Magnes et al., 1961a,b). Electrical stimulation in the low frequency range of 1 to 20 Hz to a region within the nucleus of the tractus solitarius (NTS) and to a small area adjacent to it was effective in producing EEG synchronization of the cerebral cortex. This finding raised two questions. First, what is the natural, physiological influence to the bulbar brain stem that activates the ascending inhibitory effect? And second, what is the target of the ascending influence that produces the cortical synchronization?

Answers to these questions eventually solved a serious issue raised earlier. Recall that the descending, physiological components of arousal result in positive feedback to the brain that further increases arousal and positive feedback. This is an untenable state of affairs. We know this does not happen simply because the act of awakening does not spin out of control into a grand mal seizure. The answer to the first question (What is the physiological mechanism that activates the ascending inhibitory influence?) was revealed by experiments already mentioned that were done during the 1950s (Dell et al., 1961).

Dell and his colleagues demonstrated that one of the powerful sources of input to a region of the nucleus of the tractus solitarius is the Hering branch of cranial nerve IX. This branch carries first-order afferents associated with baroceptors (pressure receptors) of the carotid sinuses. The carotid sinuses are located in the neck, one in each of the carotid arteries, which carry arterial blood to the brain. When autonomic (descending) arousal occurs, the baroceptors detect an increased blood pressure and feed signals to the brain that activate the inhibitory area of the bulbar brain stem. This, in turn, results in a counterbalancing correction to prevent excessive arousal due to the positive feedback consequences of arousal. Therefore, built into the process of descending arousal, along with elements of positive feedback that posed the positive feedback dilemma, there is at least one regulatory, dampening process.

The second question, regarding the target of the ascending inhibitory influence from the medulla, was answered by experiments of Mauro Mancia, a former student of Moruzzi, now at Milan. Mancia and his coworkers, using extra- and intracellular single-cell recordings, showed that bulbar brain stem stimulation has an inhibitory effect upon neural activity of the rostral reticular formation as well as upon diencephalic structures involved in arousal (Mancia et al., 1974). Figure 6.2 adds to Figure 6.1 the negative feedback influences from the forebrain (F) and lower brain stem (G) that serve to dampen arousal and produce slow-wave sleep.

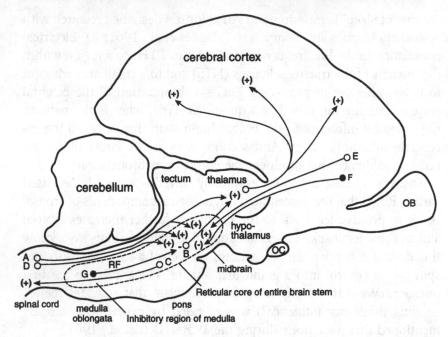

FIGURE 6.2. This figure duplicates Figure 6.1 with the addition of negative feedback from the forebrain (F) and from the inhibitory region of the lower brain stem (G). The effect of negative feedback to the rostral reticular formation is to decrease arousal. Abbreviations: OB, olfactory bulb; OC, optic chiasm; RF, reticular formation.

Until the mid-1950s, sleep was considered to be the baseline state of the brain. Bremer's view, based on observations of his *cerveau isolé* preparation, was typical: when there is a minimum of sensory input, the brain reverts to its slow-wave sleep state. This is relatively simple and has a commonsense appeal. It follows that arousal is due to sensory stimulation that abolishes sleep. The research of Moruzzi, Magoun, and Lindsley on the reticular activating system helped to explain how that worked.

The view that falling asleep is due to a reduction of sensory input changed in the 1950s with evidence that activation of certain brain structures has the capability to reduce arousal. The framework of a centrally located arousal system that can be down-regulated by the regions of the cortex and lower brain stem to produce sleep was generated mainly by Dell, Bonvallet, Moruzzi, and their colleagues. Since then, activation of other brain structures has been shown to also produce EEG synchronization and, in some cases, behavioral inhibition and sleep. That work will be described in the next chapter, followed by a discussion of sleep as a passive or an active process of cerebral deactivation.

7

EEG Synchrony and Behavioral Inhibition

The Raphé Nuclei and Serotonin

The raphé nuclei of the brain stem and the neurotransmitter serotonin have long been implicated in the production of sleep. Michel Jouvet at the University of Lyon has been the major proponent of this position. In the 1950s, Jouvet and his colleagues manipulated brain levels of serotonin in two ways. They used drugs that inhibited serotonin breakdown, thus increasing its availability at synapses, and drugs that prevented serotonin synthesis, thus decreasing its availability for synaptic release. Slow-wave sleep was enhanced when serotonin levels were increased, and insomnia was produced with reduced serotonin. However, papers appeared that presented data inconsistent with the role of serotonin in slow-wave sleep. Two such reports showed that activation of the nucleus raphé dorsalis, considered to be the major source of forebrain serotonin, failed to produce EEG synchronization and sleep, though it did produce inhibition of behavior that has been associated with slow-wave sleep (Jacobs et al., 1973; Siegel and Brownstein, 1975).

When it was determined in the 1960s that the raphé nuclei, located along the midline of the entire brain stem, were the major repository of serotonin neurons (Dahlstrom and Fuxe, 1964), Jouvet and his colleagues embarked on a program of raphé lesions. The serotonergic system of the brain was totally or partially destroyed by electrolytic lesions and by serotonin-selective neurotoxins. Barbara Jones, at McGill University in Montreal, worked with Jouvet on some of these studies and described the results of these experiments in a review paper on sleep: "Subtotal lesions . . . were associated with variable decreases in sleep, and the amount of slow-wave sleep was correlated with the percentage of destruction of the raphé nuclei and the percentage of depletion of serotonin in the forebrain" (Jones, 1989, p. 130). All of this was supportive of Jouvet's earlier work in which serotonin levels were manipulated pharmacologically. However, a severe blow to the position that serotonin and the raphé nuclei play a central role in sleep, as postulated by Jouvet, came when animals were permitted to recover from raphé lesions for extended periods of time. The loss of slow-wave sleep gradually dissipated, and the EEG and behavioral signs of sleep returned. Jones states that ". . . serotonin neurons may normally facilitate the onset of sleep but they are not essential for the occurrence of slow-wave sleep. They were, therefore, considered to be one component of a larger sleep-generating system" (Jones, 1989, p. 131).

Jouvet and his colleagues have explored the role of the raphé nuclei and serotonin as components of a larger sleep-generating system. In 1985 the Lyon lab presented evidence that the dorsal raphé nucleus, via a serotonergic projection to the hypothalamus, is involved during waking in the synthesis and accumulation of a hypothalamic sleep factor (Sallanon et al., 1985). More recently, in 1995, Jouvet and coworkers (El-Kafi, et al., 1995) showed that the dorsal raphé nucleus, *independent of the transmitter serotonin*, plays a role in sleep. Rats were pretreated with p-chlorophenylalanine (PCPA), a drug that depletes the brain of serotonin and produces insomnia. Peptides known to have sleep-inducing effects were then injected into the dorsal raphé nucleus and sleep was restored. This demonstrated that the sleep-inducing effect of these peptides upon the raphé nucleus was not mediated by serotonin.

In this context, it is important to point out a known but often ignored fact of regional neurochemistry. A nucleus that is described, for example, as serotonergic contains significant numbers of cells that synthesize and release neurotransmitters other than the transmitter that the nucleus is characterized as having. These "other" cells include not only short axon, local circuit neurons such as GABA inhibitory interneurons, but projection neurons that connect to distant regions and release their own "other" neurotransmitters. In some cases another transmitter may be colocalized within the same neurons that synthesize the principal transmitter of that nucleus; in other cases different populations of neurons within the nucleus contain the different neurotransmitters. This could account for Jouvet's finding showing a sleep role of the dorsal raphé nucleus that is mediated by a nonserotonergic pathway.

Thalamic Recruiting and Other Forms of Induced Slow Waves

An early demonstration that the thalamus played a role in sleep was given in Zurich, Switzerland, by W. R. Hess, who reported in 1944 that low frequency stimulation (6–8 pulses per second) of the medial thalamus induced behavioral sleep in the cat (Hess, 1944). In a series of papers published in 1942 and 1943, Morison and Dempsey showed that low frequency stimulation of the medial, intralaminar, and midline thalamic nuclei produced a stylized form of EEG synchronization called "recruiting waves," recorded from the cortex of the cat (Morison and Dempsey, 1942; Dempsey and Morison, 1942a,b; 1943). The first of a series of shocks to one of these thalamic nuclei evoked a barely perceptible cortical field potential. Successive shocks of the same intensity evoked progressively larger responses of negative polarity to a peak amplitude that then gradually subsided in amplitude. Figure 7.1

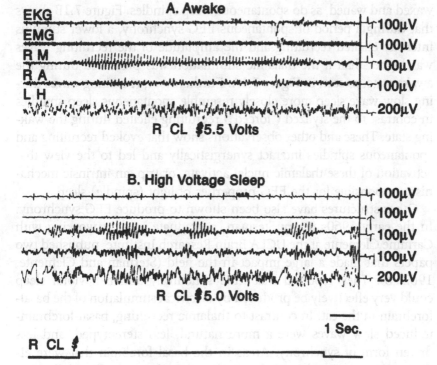

FIGURE 7.1. Recruiting waves recorded from cat cerebral cortex in response to electrical stimulation of 7.5 pulses per second delivered to nucleus centralis lateralis (CL), one of the intralaminar nuclei of the thalamus. Stimulation was presented during the awake state (A) and during slow-wave (high voltage) sleep (B). The five recording channels show, from top to bottom, the electrocardiogram (EKG), electromyogram (EMG), and EEG traces from the right frontal cortex (RM), right parietal cortex (RA), and the left hippocampus (LH). Thalamic recruiting is more readily recorded from frontal than from other cortical regions. The text discusses differences in the recruiting response seen during waking (A) and sleep (B). The line at the bottom of the figure marks the onset and offset of thalamic stimulation. The symbol S with an arrow through it signifies electrical stimulation. (Adapted with permission from Siegel, 1967, p. 143. Copyright © 1967 by the American Psychological Association.)

illustrates recruiting waves recorded from the cerebral cortex of a cat. The envelope around this waveform is spindle shaped. The significance of recruiting waves is that they resemble spontaneous spindle waves, seen in the early stages of slow-wave sleep, which also are predominantly negative in polarity.

The term "recruiting" was attached to these thalamically induced slow waves based on the notion that each successive shock to the thalamus recruits more thalamocortical neurons into a firing pattern that generates increasingly larger field potentials at the cortex. If the stimulating voltage was considerably higher than the threshold necessary to produce the recruiting response, the negative waves remained at maximal amplitude. If a threshold level of voltage was used, the recruiting waves waxed and waned, as do spontaneous sleep spindles. Figure 7.1B shows that during a period of spontaneous EEG synchrony, a lower stimulus intensity elicited recruiting and the amplitude of the recruiting waves was higher than if the thalamic stimulation was presented during the awake state, as seen in Figure 7.1A. Furthermore, recruiting waves during slow-wave sleep more closely resemble spontaneous sleep spindles in contrast to the stylized ("forced") recruiting elicited during the waking state. These and other observations show that evoked recruiting and spontaneous spindles interact synergistically and led to the view that activation of these thalamic nuclei activate, or trip, an intrinsic mechanism responsible for the EEG properties of light (spindle) sleep.

Other structures have also been shown to produce EEG synchrony. In the early 1960s, Barry Sterman, a graduate student working with Carmine Clemente at the UCLA Brain Research Institute, published two papers that made a large impact in the field (Sterman and Clemente, 1962a,b). They reported that EEG synchrony and behavioral sleep could very effectively be produced by electrical stimulation of the basal-forebrain in the cat. In contrast to thalamic recruiting, basal-forebrain-induced slow waves were a more natural, less stereotyped, and less driven form of synchrony. Areas in the basal forebrain that were effective in producing this synchrony were found to extend caudally from the rostral basal forebrain to the preoptic area and anterior hypothalamic region. The three areas are anatomically contiguous and, for the sake of simplicity, will be referred to as the extended basal forebrain. Figure 7.2 illustrates the EEG synchrony reported by Sterman and Clemente (1962a) and the area of the forebrain they stimulated to produce the synchrony. Figure 7.3 shows the locations of the rostral basal forebrain, preoptic, and anterior hypothalamic areas (the extended basal forebrain) relative to each other and to other parts of the brain.

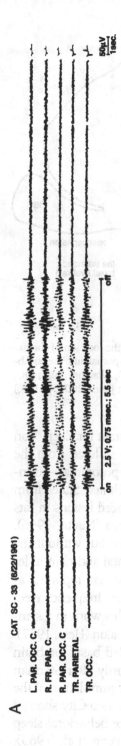

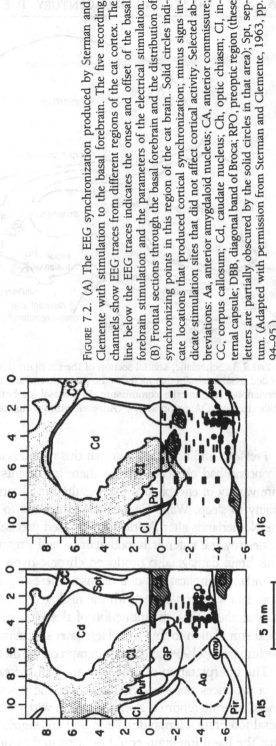

FIGURE 7.2. (A) The EEG synchronization produced by Sterman and Clemente with stimulation to the basal forebrain. The five recording channels show EEG traces from different regions of the cat cortex. The line below the EEG traces indicates the onset and offset of the basal forebrain stimulation and the parameters of the electrical stimulation. (B) Frontal sections through the basal forebrain and the distribution of synchronizing points in this region of the cat brain. Solid circles indicate locations that produced cortical synchronization; minus signs indicate stimulation sites that did not affect cortical activity. Selected abbreviations: Aa, anterior amygdaloid nucleus; CA, anterior commissure; CC, corpus callosum; Cd, caudate nucleus; Ch, optic chiasm; CI, internal capsule; DBB, diagonal band of Broca; RPO, preoptic region (these letters are partially obscured by the solid circles in that area); Spt, septum. (Adapted with permission from Sterman and Clemente, 1963, pp. 94–95.)

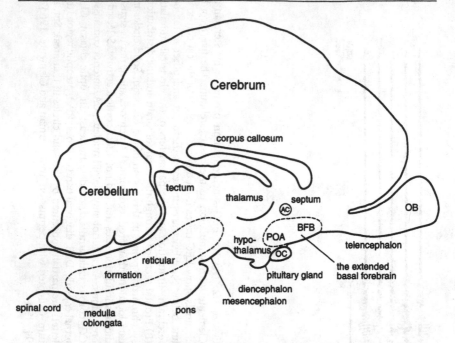

FIGURE 7.3. Schematic, sagittal section of the cat brain showing the location of the extended basal forebrain that includes the preoptic and anterior hypothalamic areas. Abbreviations: AC, anterior commissure; BFB, basal forebrain; OB, olfactory bulb; OC, optic chiasm; POA, preoptic area. (Adapted with permission from Jones, 2000, p. 138.)

Previous findings suggested that this region is important for sleep. Von Economo had observed lesions there in patients who suffered from the rare variety of encephalitis that resulted in chronic hyposomnia—an inability to sleep. Walle Nauta, who emigrated to the United States from the Netherlands after World War II, used precisely placed lesions in rats to destroy the anterior hypothalamus and preoptic area (Nauta, 1946). This brain damage also produced chronic hyposomnia, but without the attendant complications of a disease. These experimental and clinical lesion data were consistent with the finding of Hess, who had reported, as early as the 1930s, the induction of sleep in cats by low frequency electrical stimulation to this area. Hess later summarized this work in a symposium paper (1954) and a monograph on the diencephalon (Hess, 1957).

The interpretation of how activation of the extended basal forebrain area produces sleep has changed over the years. Shortly after Sterman and Clemente reported sleep induction with electrical stimulation of the basal forebrain, Hernandez-Peon and colleagues in Mexico City showed that the basal forebrain could be activated to produce behavioral sleep by the local application of acetylcholine (Hernandez-Peon et al., 1963).

Cells of this region have receptor sites that bind acetylcholine and are described, therefore, as cholinoceptive. In addition, it has been shown that the basal forebrain contains cells that are cholinergic; that is, they synthesize acetylcholine and release that transmitter at their terminals. So these cells are both cholinoceptive and cholinergic.

In the early version of how this system works, the basal forebrain cholinergic neurons were thought to project to a number of regions, the primary target being the cerebral cortex where the cholinergic influence was the production of EEG slow waves and behavioral sleep. There was also the suggestion that the basal forebrain sleep effect was mediated by a projection to the thalamus that, in turn, contributed to the EEG synchrony and sleep.

After much research, the story has changed considerably. The view that the basal forebrain contains neurons that are cholinoceptive and cholinergic has held up. As described in Chapter 5, on arousal, the cholinergic input to these cells is from the other major cholinergic population in the brain, the peribrachial cells of the brain stem reticular formation (see Figures 5.1 and 5.2). Recall that the ventral pathway from the cholinergic reticular formation has two targets. One is the posterior lateral hypothalamus, which produces arousal via a histamine projection to the cortex; the other target is the basal forebrain. There is considerable evidence now that the cholinergic cells of the basal forebrain do project to the cerebral cortex; however, this cholinergic influence produces *arousal,* not sleep. That is a large change from the early view! How, then, do we explain the sleep produced by electrical stimulation to the basal forebrain?

The view articulated by Jones (1993) and Saper et al. (1997) is that the basal forebrain is important in the generation of sleep, but not via its cholinergic cells. In the mid-1980s it was discovered that comingled among cholinergic neurons of the basal forebrain are cells that synthesize and release γ-aminobutyric acid (GABA). It has been known for a long time that GABAergic neurons are ubiquitous within the central nervous system as local circuit, short-axoned, inhibitory interneurons that exert inhibitory control functions within their local areas. In recent years it has been discovered that not all GABAergic neurons are local circuit interneurons. There are also GABAergic projection neurons. GABA neurons in the extended basal forebrain project to a number of regions. One prominent target is the tuberomammillary nucleus of the posterior hypothalamus. This histaminergic nucleus is a component of the ventral cortical arousal pathway described in Chapter 5 on the neural pathways for arousal. The conclusion reached is that the extended basal forebrain area is a sleep-inducing area by virtue of exerting an inhibitory effect on at least one major component of the arousal system. This is an active deactivation process to produce sleep. Supporting this conclusion is the finding that lesions of the extended basal forebrain area that produce a loss of sleep (Szymusiak and McGinty, 1986), as first shown by von Economo, can be reversed by a local injection into the posterior hypothalamus of muscimol, a drug that mimics the effects of GABA (Sallanon et al., 1989).

Another important target of the GABAergic cells of the extended basal forebrain area, also via the medial forebrain bundle, is the peribrachial nuclei of the reticular formation. The peribrachial nuclei, as described earler, have as one of their targets the basal forebrain. This is yet another instance of what Dell and his colleagues explored, namely, a regulatory negative feedback as a consequence of activating the reticular activating system. We expect, however, that the sleep-inducing system must do more than function as a negative feedback to regulate arousal level. We expect that it should also be activated at the end of the day to produce normal nighttime sleep. There is evidence that a chemical, adenosine, is synthesized during the increased, sustained neural activity associated with arousal and acts to inhibit the cells of the cholinergic arousal system. Inhibition of arousal leads to sleep. This material will be covered in more detail in Chapter 9, Sleep Factors.

Two other forebrain structures have been shown to be effective in producing EEG synchrony. Peñaloza-Rojas et al. (1964) and Clemente (1968) reported that stimulation to the orbital gyrus of the frontal cortex of cats produced EEG synchronization and behavioral sleep, similar to what Sterman and Clemente (1962a,b) had shown with basal forebrain stimulation. The orbital gyrus is located at the ventral aspect of the frontal cortex and is anatomically and functionally linked with the basal forebrain area described earlier.

The caudate nucleus is the other forebrain structure involved in EEG synchrony. Nathaniel Buchwald and his colleagues at UCLA pioneered research on this aspect of its function (Buchwald et al., 1961a–c; Heuser et al., 1961). The caudate nucleus, one of the subcortical nuclei of the basal ganglia located in the forebrain, has traditionally been associated with the control of movement. However, there is evidence that the caudate nucleus and other structures of the basal ganglia also have sensory, motivational, and cognitive functions (Schneider and Lidsky, 1987; Percheron et al., 1994).

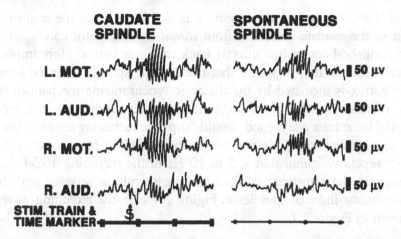

FIGURE 7.4. Caudate-induced and spontaneous spindle bursts recorded from the same electrodes in a cat. The traces on the left show caudate-induced spindles that were recorded during waking. The traces on the right show spontaneous spindle waves recorded during natural slow-wave sleep. The occurrence of the stimulus shock to the caudate nucleus is marked by the symbol **S** with an arrow through it. The time marker indicates seconds. The duration of the spindle discharge is about one second. Abbreviations: L, left; R, right; MOT, motor (pericruciate) cortex; AUD, auditory (ectosylvian) cortex. (Siegel, 1987, p. 184. Reproduced with permission from J.S. Schneider and T.I. Lidsky, *Basal Ganglia and Behavior*, ISBN 0-920887-13-9 and ISBN 3-456-81539-5. © 1987 by Hans Huber Publishers/Toronto. Lewiston, NY · Bern · Stuttgart.)

Buchwald's group showed that the caudate nucleus has synchronogenic properties associated with its inhibitory effects on behavior. A single electrical pulse to the caudate nucleus evoked a field potential recorded from the frontal cortex followed by an afterdischarge that is very similar to a spontaneous spindle burst seen in slow-wave sleep. Figure 7.4 shows caudate-induced spindles and spontaneous spindle bursts recorded from the same cortical electrodes.

Heuser et al. (1961) described an important characteristic about the "caudate spindle," as they referred to it. When the fiber pathway between the caudate nucleus and the thalamus was severed, stimulation to the caudate nucleus still produced the initial field potential component recorded from the frontal cortex, but the spindle after discharge was abolished. The early component of the field potential recorded from the cortex is considered to be antidromically* generated because

---

*An antidromic response is due to neural activity that is artificially generated at axonal terminals or in the axon, then conducted back to the cell body and recorded there. The artificially created action potential is conducted in the opposite direction to orthodromic (normal) axonal activity.

it is of short latency. Furthermore, it is well known that the major input to the caudate nucleus is from motor and premotor cortex, with no reciprocal connection directly back to motor cortex. More importantly, this finding suggested that the caudate spindle recorded from the cortex is mediated by the thalamic synchronizing mechanism of the recruiting nuclei. Thus, single shocks to the caudate nucleus, separated by at least one second, would "trip" the recruiting nuclei to produce a natural spindle burst to each shock. This is in contrast to direct, repetitive stimulation at 5 to 10 Hz of the recruiting nuclei that generates the somewhat artificial-looking recruiting waves. Compare the caudate-induced spindles in Figure 7.4 with the recruiting waves shown in Figure 7.1.

Behavioral Inhibition

In the fourth paper of their series of papers in "the *EEG Journal*," Buchwald and his colleagues reported findings with implications for the behavioral significance, other than sleep, of the caudate-induced spindle and of EEG synchrony in general (Buchwald et al., 1961b). Cats with chronically implanted electrodes were trained to press a lever to receive a small amount of cream. When the caudate nucleus was stimulated with parameters that produced a spindle discharge, the lever pressing slowed and, at higher stimulation intensities, completely ceased. Buchwald and colleagues demonstrated behavioral inhibition as a behavioral concomitant of the caudate spindle.

It has since been shown that in addition to the caudate nucleus, other brain structures capable of generating EEG synchronization also produce behavioral inhibition. In the early 1970s my students and I (Lineberry and Siegel, 1971; Plumer and Siegel, 1973; Siegel and Wang, 1974) described behavioral findings from chronically prepared cats with electrodes in the caudate nucleus, basal forebrain, and orbital cortex. When these structures were stimulated with threshold currents to produce cortical synchrony, a cessation or slowing of goal-directed behavior was observed. This may be interpreted as an interference with an ongoing motor sequence by an artificial stimulation to a brain structure. This occurs when electrical stimulation is intense enough to produce an obvious motor freezing to the exclusion of all other behavior. However, when the behavior was of a moderate nature, such as a cat quietly lapping milk from a dish, in contrast to an extreme form of behavior, such as aggressive attack, and the synchronogenic stimulation was delivered at threshold or near threshold intensity for the EEG synchrony, the cat would stop drinking and sit down. At the termination of stimulation, the cat would resume drinking. If the stimulation period was increased to about a minute, the cat would often sit with its eyes closed during the period of induced slow waves.

These findings of behavioral inhibition and EEG synchrony may be related to a phenomenon reported by Pavlov (1923), who showed a link between behavioral inhibition and sleep. One of the variants of the classical conditioning paradigm that Pavlov used with his dogs was delayed conditioning. This conditioning paradigm is represented in Figure 7.5. In this procedure the conditioned stimulus (CS) (often a sound) was turned on for a given duration (e.g., 1–2 minutes). At the end of this period the unconditioned stimulus (US), food, was delivered to the food-deprived dog. The unconditioned response (UR) to the food was salivation. After a number of trials, but still early in the training period, the CS triggered the conditioned response (CR) of salivation upon presentation of the CS. With continued training, which taught the dog very well that food would be delivered only when the delay period was over, the animal inhibited the impulse to respond at CS onset and exhibited the CR at the end of the delay period. Pavlov noted that during the delay period when the inhibitory process had been well established, the dog became drowsy and would often slump into its restraining harness and sleep—and emit an occasional snore. A re-

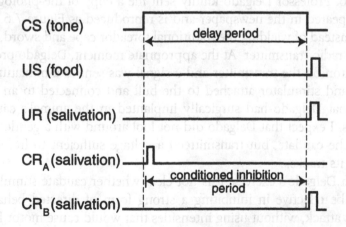

FIGURE 7.5. The Pavlovian delayed conditioning paradigm. Abbreviations: CS, conditioned stimulus; US, unconditioned stimulus; UR, unconditioned response; CR$_A$, the conditioned response that occurs at the start of the delay period and is seen near the beginning of the training experience; CR$_B$, the conditioned response that occurs at the end of the conditioned inhibition period and is seen after considerable training trials have occurred.

vealing observation by Pavlov was that if a sudden novel stimulus occurred during the delay period, the conditioned salivation was triggered. This implied that an active process of inhibition was in place
and that the distracting stimulus had interfered with the inhibitory
process and thus released the response.

It appears that a state of internal inhibition produced by this form
of classical conditioning involves the activation of a synchronogenic
process similar to that of normal sleep and similar to that activated by
gentle electrical pulsing of the brain structures just described. In all
three cases, the EEG synchrony reflects an inhibition of the cortex that
interferes with information processing and with the elicitation of goal-
directed behavior. This is quite plausible. For example, when a state
of drowsiness overtakes one, information processing and most forms
of directed behavior tend to cease. When the brain is thrown into a
state of EEG synchrony by electrical stimulation or by a behavioral con-
tingency (e.g., Pavlov's delayed conditioning), the same occurs. It may
be concluded that synchronogenic circuits of the brain are not limited
to controlling normal nighttime sleep and drowsiness, but also are activated in mechanisms of behavioral inhibition.

A dramatic demonstration of caudate-induced inhibition was reported in the *New York Times* in 1965. Jose Delgado, a prominent neu-
roscientist then at Yale University, was photographed in a bullring in
Cordoba, Spain, with a large bull stopped midstream in an attack upon
Delgado. Professor Delgado kindly sent me a copy of the photograph
that appeared in the newspaper and is reproduced as Figure 7.6. Delgado, instead of wielding the traditional toreador cape and sword, held
a small radio transmitter. At the appropriate moment, Delgado pressed
the button on the transmitter and a signal was sent to a miniature re-
ceiver and stimulator attached to the bull and connected to an electrode that Delgado had surgically implanted in the animal's caudate
nucleus. I expect that Delgado did not fool around with a gentle pulsing of the caudate, but transmitteed a voltage sufficient to freeze the
bull in its tracks.

From Delgado's exercise it is not clear whether caudate stimulation
would be effective in inhibiting a strong form of directed behavior,
such as attack, without using intensities that would cause motor freezing. My students and I explored this issue, albeit not as exotically as
Delgado did. We implanted electrodes into the caudate nucleus and in
a region of the hypothalamus that, when electrically stimulated, produced attack behavior in cats (Wasman and Flynn, 1962). We separately determined the hypothalamic stimulation parameters that pro-

FIGURE 7.6. Jose Delgado in a bull ring, inhibiting an attack from a bull by remote control stimulation of the caudate nucleus. (Photograph kindly provided by Professor Delgado and reproduced with permission.)

duced attack behavior and the threshold intensities of caudate stimulation that produced EEG synchrony. When the hypothalamically elicited attack was of a predatory nature (e.g., a cat stalking an anesthetized rat), gentle pulsing of the caudate nucleus at or slightly above the threshold for eliciting EEG synchrony was adequate to inhibit the attack. We showed that attack behavior could be inhibited without having to resort to intensities that would produce freezing behavior (Plumer and Siegel, 1973). Figure 7.7 shows the rather natural look of the EEG synchronization produced by the caudate nucleus stimulation used in that experiment.

Sleep as Passive or Active Deactivation

An interesting question was posed when it became known that there are synchronogenic and somnogenic sites distributed throughout the brain from the caudal medulla to the cerebral cortex: What are the neural mechanisms by which these synchronogenic effects are produced? Some of this material has been covered. In this section I will deal with a related question: Is sleep due to a passive process of sensory deafferentation or an active process of reticular deactivation? Bremer's po-

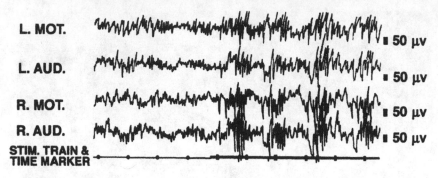

FIGURE 7.7. Caudate-induced EEG synchronization recorded from the cat cortex to caudate stimulation at a rate of one shock every 1.5 seconds. The bottom trace shows 1-per-second time marks and, where the trace thickens, the period during which caudate stimulation was presented. Abbreviations: L, left; R, right; MOT, motor (pericruciate) cortex; AUD, auditory (ectosylvian) cortex.

sition for a number of years was that sleep is due to a removal of sensory afferentation of the cerebrum. His early view was that the *encéphale isolé* preparation (an isolated head and brain) showed both sleep and waking states because the cranial nerves were intact and allowed for sensory input to the cerebrum and thus waking. He interpreted the *cerveau isolé* preparation (isolated cerebrum) as showing only sleep because it was mainly deafferented and devoid of inputs required for waking. Thus, sleep for Bremer was a passive process (Bremer, 1935, 1938). However, between the late 1960s and early 1970s Bremer altered his position on this issue.

After the publication by Moruzzi and Magoun in 1949 and the lesion studies of Lindsley et al. of 1949 and 1950, Bremer had accepted the role of the mesencephalic reticular formation as a central structure mediating arousal. In a 1970 paper, he stated that he had been persuaded by the Sterman and Clemente and the Pisa experiments that EEG synchronization and sleep can be produced by stimulation of the basal forebrain and the bulbar brain stem, respectively. With evidence mounting and pointing to the possibility that sleep may be due to an active deactivation of the reticular formation, Bremer conducted an experiment that would be telling with respect to that position.

This study was designed to determine whether the induction of synchrony and sleep was mediated by an inhibitory effect on the rostral reticular formation. Using his *encéphale isolé* preparation, Bremer placed a stimulating electrode in the basal forebrain and a recording electrode in the reticular formation to detect field potentials evoked by the basal forebrain stimulation. Shortly after a single shock to the basal forebrain, a brief negative-going field potential followed by a larger and longer duration positive wave and another negativity were recorded from the reticular formation. The dominant positive wave was interpreted as a reflection of synaptic inhibition of cells in the reticular formation due to stimulation of the basal forebrain. The initial negativity was seen as a field potential component due to an initial burst of single cell activity elicited by the forebrain shock, and the late slow negative wave after the inhibitory period was viewed as a postinhibitory rebound at the cellular level. Bremer also delivered paired shocks to the basal forebrain. A pronounced enhancement of the slow positive wave followed the second shock; this was interpreted as reflecting a process of temporal summation of inhibitory postsynaptic potentials. These data provided proof to Bremer that basal forebrain stimulation produces EEG synchronization and sleep by inhibiting neural activity in the rostral reticular formation. In this paper entitled Preoptic Hypnogenic Focus and Mesencephalic Reticular Formation (Bremer, 1970), he acknowledged too that the synchronogenic properties of the bulbar brain stem could also be due to deactivation of the ARAS.

In this context, I recall a paper I presented in 1971 at the First International Congress of the Association for the Psychophysiological Study of Sleep (APSS) in Bruges, Belgium, at which Bremer was present. (Bremer and Kleitman shared the honorary presidency of the congress.) I described data recently collected by me and a graduate student, Charles Lineberry, in which the orbital cortex, basal forebrain, and caudate nucleus in chronically prepared cats had been stimulated at parameters that produced EEG synchronization and behavioral inhibition (Siegel and Lineberry, 1968; Lineberry and Siegel, 1971). Initially, during behavioral sessions, we determined the threshold currents necessary to produce the EEG and behavioral effects. In subsequent sessions with the same cats, single-cell action potentials were recorded extracellularly from neurons of the mesencephalic reticular formation as the synchronogenic structures were stimulated at the parameters that produced the EEG and behavioral inhibitory effects. I presented slides showing that at threshold currents for the electrographic and behavioral effects, cells in the reticular formation showed an altered pattern of firing. In many cases, this involved a short latency, short burst of spikes followed by a decreased firing rate or a complete cessation of unit activity. Figure 7.8 shows oscilloscope traces of unit activity recorded from a single cell of the reticular formation that was profoundly inhibited by caudate stimulation that also produced EEG synchrony at the cortex. In some cells the inhibitory pe-

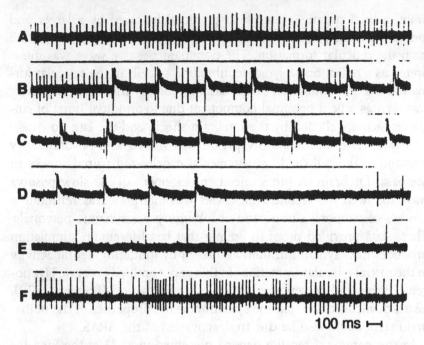

100 ms ⊢—⊣

FIGURE 7.8. A cell in the mesencephalic reticular formation that was completely in-hibited by 3-per-second caudate stimulation that also produced EEG synchronization at the cortex. (A) Oscilloscope trace showing a spontaneous prestimulus firing rate of about 30 spikes per second. (B) The first and subsequent shocks of a one-minute pe-riod of stimulation. Continuous in time with (B), trace (C) shows the caudate-induced decrease in the firing rate of the cell. Each large potential is an artifact from the cau-date stimulus. The end of the stimulation period is seen in trace (D). The cell did not resume firing again until about a second after the stimulation had stopped. Continu-ous in time with (D), trace (E) shows the very gradual return of neural activity. The last trace, (F), started 50 seconds after the end of caudate stimulation, shows that the cell had returned to its spontaneous firing rate by then. (Reproduced with permission from Siegel and Lineberry, 1968, p. 449.)

riod was followed by a postinhibitory rebound of spike activity. This pattern of single-cell activity perfectly reflected what Bremer had ob-served at the field potential level and confirmed his interpretation of the field potential data.

After my talk, Bremer approached me and said that the combined evidence of basal forebrain-produced EEG synchrony and sleep and my direct evidence that when those EEG and behavioral effects are pro-duced, cells of the ascending reticular activating system are inhibited, had persuaded him that there are processes in the brain that initiate and maintain sleep by actively inhibiting the reticular arousal system. Bremer, however, commented that he still held to his position that the passive process of sensory removal was also relevant for sleep and that the two mechanisms were not mutually exclusive. Subsequently, in a contributed chapter entitled Historical Development of Ideas on Sleep, Bremer wrote "I still believe that the passive and active determinants of sleep—more precisely of ordinary slow-wave sleep—are complementary notions which should not be regarded as conceptually opposed" (Bremer, 1974, p. 7). I expect that most workers in the field of sleep would agree with this position.

Since the early 1970s, evidence has accumulated that the basal forebrain, including the preoptic area and the anterior hypothalamus, is the major region of the forebrain responsible for the induction of sleep. This is accomplished by a deactivation of the arousal system at the level of the posterior hypothalamus and the rostral brain stem. It is interesting to note that this view is not far off from that published by von Economo in 1924.

From the work described thus far, an emerging conclusion is that sleep is produced by structures distributed throughout the brain that exert inhibitory influences upon a centrally located arousal system. The distribution of sleep-generating structures ranges from the caudal brain stem to the cerebral cortex; arousal structures are limited to the rostral brain stem and adjacent structures of the thalamus and posterior hypothalamus and the basal forebrain.

8

Cellular Mechanisms and Neural Circuits That Produce Sleep

Recently developed techniques in electrophysiology have yielded new insights into the cellular mechanisms that produce EEG synchrony and sleep. These advances include the use of the brain slice preparation in which a slice of brain tissue about 400 μm thick is placed in a medium that provides the necessary oxygen and nutrients to maintain cell viability for a day or more. These cells are devoid of most of their afferent influences, and their intrinsic properties can be studied with intracellular electrodes. Moreover, the bath medium and intracellular contents can be manipulated to investigate the roles of various neurotransmitters, receptor types, and their associated ion channels. This has been accomplished by the addition to the bath medium of specific receptor blockers, enzymes that break down and inactivate certain neurotransmitters, and altered concentrations of ions such as sodium, potassium, chloride, and calcium. The brain slice preparation has contributed much to our understanding of the cellular mechanisms that underlie the EEG recordings taken from intact subjects during sleep and waking.

Two investigators have been prominent in this effort: Mircea Steriade at Lavale University in Quebec and David McCormick at Yale University. Working independently from each other, they have developed a common view, well documented by experimental data, about thalamic control of the EEG states seen in sleep and waking. Much of their work and the related work of others is described in a book by Steriade and McCarley (1990) and in a review paper by McCormick and Bal (1997).

The role of the reticular nucleus of the thalamus is central to their view of slow-wave sleep. The reticular nucleus is unique in that, un-

79

like all other thalamic nuclei, it does not project to the cerebral cortex. The reticular nucleus is also strategically located. The nucleus comprises a relatively thin shell of cells that overlies the thalamus at its anterior and lateral aspects, so that fibers originating in the thalamus and projecting to the cerebral cortex pass through the reticular nucleus. Similarly, corticothalamic fibers do the same. In addition, the axons that pass through the reticular nucleus provide collateral fibers that synapse upon cells of the reticular nucleus. In this way, the reticular nucleus monitors the two-way traffic between the thalamus and cortex. Now, what does it do with this information? According to Steriade and to McCormick, the reticular nucleus projects to the other thalamic nuclei and plays a major role in the thalamic regulation of the arousal state of the cortex.

Neurons of the reticular nucleus have an intrinsic property that is characteristic of pacemaker cells. During slow-wave sleep, when influences present during arousal are absent and cells of the reticular nucleus are left to their own intrinsic devices, these cells exhibit their pacemaker properties. With no external driving influence, the resting membrane potentials of reticular nucleus cells undergo phasic shifts between states of depolarization and hyperpolarizaton. During these shifts of membrane polarization levels, the membrane channels for calcium, sodium, and potassium are activated (opened) and inactivated (closed) in complex patterns to produce a characteristic firing pattern of these cells (Steriade and Deschenes, 1984). When cells of the reticular nucleus reach the depolarization threshold for firing, they generate bursts of spike action potentials. The depolarized resting membrane level gradually dissipates to a value below the threshold for action potentials, and the cells enter a hyperpolarization phase during which they are quiet. The inhibitory phase also gradually dissipates to the threshold level of generating the next burst of action potentials, and so on. The kinetics and intrinsic properties of these cells result in rhythmic bursts at frequencies of 0.5 to 12 Hz.

Projection neurons of the reticular nucleus are predominantly GABAergic. The inhibitory transmitter γ-aminobutyric acid exerts a strong inhibitory drive upon thalamocortical neurons to which the GABAergic neurons project. The inhibitory postsynaptic potentials (IPSPs) generated by this inhibitory drive produce postinhibitory rebounds (depolarization) that trigger a train of spike potentials in thalamic neurons similar to those seen in reticular nucleus neurons. As a consequence, thalamocortical neurons, which utilize an excitatory amino acid (probably glutamate) as their transmitter, drive their cortical targets into a rhythmic bursting pattern characteristic of slow-wave sleep. The thalamocortical cells pass through the reticular nucleus on their way to the cortex and provide axonal collaterals that produce an excitatory drive to cells of the reticular nucleus. This generates excitatory postsynaptic potentials (EPSPs) that reinforce the bursting pattern that initiated the process. In addition, the cortical cells of layer 4, which receive the afferent projections from the thalamus, are thrown into a bursting pattern by their excitatory input from the thalamus. The layer 4 cells exert, in turn, a short latency excitatory drive upon deeper layer corticothalamic cells. These deep cortical cells have reciprocal glutamatergic (excitatory) connections with the thalamic nuclei from which the cortical layer 4 cells had received an input. The corticothalamic neurons also provide collateral axons to the reticular nucleus, so that the reticular nucleus is reinforced in its firing pattern yet again. Figure 8.1 is a schematic representation designed to help understand the complex circuit just described.

The consequence of all this is that during slow-wave sleep the cells of the thalamus and cortex are strongly occupied with an intrinsically programmed pattern of rhythmic firing that is inimical to the processing of sensory information. As a result, the neuronal firing patterns associated with the coding of sensory messages are not transmitted through the thalamus. Similarly, the complex patterns of neural activity that support cognitive processes are prevented from occurring. That is perfectly descriptive of what occurs in slow-wave sleep: we are oblivious to sensory events, and cognitive processing is absent. In addition, these data explain the cellular basis of the EEG patterns seen in sleep referred to as synchronization. These wave patterns are in the spindle frequency range of 8 to 12 Hz during light sleep and the delta range of 0.5 to 4 Hz in deep sleep. Both these EEG bandwidths are generated in the dendritic fields of cortical cells by the synchronous discharge of thalamic neurons that are intrinsically programmed to burst in the frequency range of 0.5 to 12 Hz. Neural activity in this frequency range is paced and reinforced by the circuit already described in which the thalamic reticular nucleus plays a dominant role during slow-wave sleep.

Steriade and McCormick also address the issue of the transition from sleep to waking. Essentially, what is the mechanism that terminates sleep? It involves a number of neural pathways and neurotransmitters. The rostral brain stem reticular formation is the structure central to arousal. That view of arousal has not changed basically since Moruzzi and Magoun advanced it over 50 years ago. However, our understanding of the detailed process by which it accomplishes its arousal

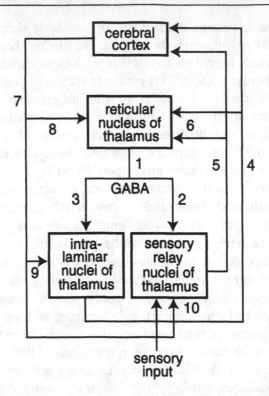

FIGURE 8.1. Role of the thalamic reticular nucleus and thalamocortical and corticothalamic circuits that regulate the state of the cortex during slow-wave sleep. Rhythmic firing of GABA inhibitory neurons of the reticular nucleus (1) drives sensory relay nuclei of the thalamus (2) into a rhythmic pattern that overrides the incoming sensory message. Rhythmic driving of the intralaminar nuclei (3), with their widespread cortical projections, produces rhythmic synchronization of the cortical EEG (4). Similarly, the rhythmic activity and output of the sensory relay nuclei (5) contribute to the cortical synchronization. Collateral axons from the intralaminar and sensory nuclei synapse upon cells of the reticular nucleus (6) to reinforce their rhythmic activity. Corticothalamic feedback from a synchronized cortex (7) also reinforces the synchronized activity of the reticular nucleus (8) and other thalamic nuclei (9 and 10).

effect has changed. The thick black arrows going from the reticular formation and thalamus to the cortex (see Figure 4.6) have been replaced by a complex of neural circuits and multiple transmitters. Regions of the pontine and mesencephalic brain stem that include the areas stimulated by Moruzzi and Magoun and others to produce cortical arousal have since been shown to contain populations of cholinergic, noradrenergic, serotonergic, and dopaminergic neurons. As described in Chapter 5, on arousal, cholinergic cells reside in two nuclei, collectively called the peribrachial nuclei. The major source of noradrenergic neurons is the locus coeruleus. Also involved in arousal and the modulation of arousal are serotonin from the raphé nuclei of the rostral brain stem, dopamine from the ventral tegmental area of the mesencephalon, histamine and hypocretin from the posterior hypothalamus, and acetylcholine from the basal forebrain. These nuclei and their neurotransmitters exert their influences upon the cortex by direct projections and indirectly by affecting the thalamus or other subcortical structures that, in turn, project to the cortex.

The transition from sleep to waking occurs when thalamocortical neurons, under the influence of certain of these neurotransmitters, undergo a slow depolarization that inhibits the rhythmic bursting pattern of neural activity and brings the membrane potential to the threshold for a single spike, nonbursting firing pattern. Under this condition, these cells fire in the frequency range of 30 to 60 Hz. A single spike firing pattern of 30 to 60 Hz is what thalamocortical neurons display during waking when they transmit the complex firing patterns associated with sensory information. Dendritic postsynaptic potentials of cortical cells firing in this frequency range will generate an EEG rhythm of 20 to 40 Hz. This is the bandwidth of beta waves seen in the waking EEG (LVFA). As more of these transmitters are released in the transition to waking, this firing pattern increasingly dominates the rhythmic spike bursting seen in slow-wave sleep.

The GABAergic projection neurons of the thalamic reticular nucleus respond to these transmitters in a fashion similar to that of the thalamocortical neurons. An exception to this is the response of reticular nucleus cells to acetylcholine from the peribrachial nuclei of the reticular formation arousal system. The thalamic reticular nucleus cells possess cholinergic receptors of a type (M2 muscarinic receptors) that cause the opening of potassium channels, thus resulting in membrane hyperpolarization and an inhibition of firing (Puoliväli et al., 1998). Therefore, during waking, thalamic reticular neurons either are inhibited by a cholinergic input or fire at a fairly steady nonbursting rate in response to other transmitters that impinge upon them. Either way, their thalamocortical target cells are no longer paced to fire in a bursting pattern and are free to fire in a single spike, information transfer mode. Figure 8.2 schematically shows the role of the peribrachial nuclei in their influence over the thalamic reticular and sensory relay nuclei during waking.

Steriade and his colleagues (1996b) show that the high frequency EEG oscillations of 30 to 40 Hz are synchronized not only within cortical areas, but also among thalamic nuclei as well as between thalamus and cortex. Steriade et al. (1996a) speculate that these oscillations

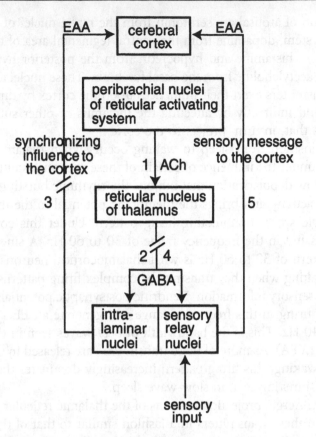

FIGURE 8.2. Role of the peribrachial nuclei during waking. During waking cells of the peribrachial nuclei of the reticular activating system exert a cholinergic (ACh) influence (1) on the thalamic reticular nucleus that inhibits the rhythmic bursting activity of the GABAergic cells of the reticular nucleus. This removes the rhythmic driving of the intralaminar nuclei (2), an event that, in turn, (3) removes a major synchronizing influence to the cortex from the intralaminar nuclei. The sensory relay nuclei of the thalamus, also freed of their rhythmic driving from the reticular nucleus (4), are now available to be occupied by sensory inputs arriving from receptors and can relay the patterns of neural activity that carry sensory information to the cortex (5). The neurotransmitter of thalamocortical neurons is an excitatory amino acid (EAA).

represent synchronous synaptic potentials among large populations of neurons engaged in neural circuits for processing external sensory stimuli during waking and for internally generated cortical activities during REM sleep.

The end result of this state of affairs is that during waking, the thalamus is freed from its intrinsic bursting pattern and is capable of transmitting sensory information to the cortex. The cortex at this time exhibits a pattern of neural activity appropriate to support the processing of the received information.

9

Sleep Factors

It is well known that after you have stayed awake for a number of hours you will get tired and sleepy. You may fight the urge to close your eyes and sleep, but with time the pull toward doing so increases and the desire to release into sleep becomes overpowering. Having done all-night experiments in which it is imperative to remain not only awake but also alert, I can personally attest to the difficulty of this task between 2 A.M. and 6 A.M. It is as if something—let's call it a chemical factor or a hormone*—builds up and causes this effect, and the longer you stay awake the more this chemical accumulates, making the effort of staying awake more difficult. Then when one eventually does sleep, this chemical dissipates, and likewise the drive to sleep.

This view of sleep has been investigated in a number of ways. The earliest approach was to remove from animals deprived of sleep a portion of blood or cerebrospinal fluid (CSF)—or brain tissue itself—and inject it into animals not so deprived. The assumption was that whatever this factor is would accumulate in the fluids or brain tissue of animals that are kept awake and could affect the brains of recipient animals, thus causing sleep. Conversely, during sleep the substance would be metabolized and inactivated and the animal would awaken.

The first studies of this nature were done independently in Japan and France during the early years of the twentieth century. Both projects used dogs. As described by Inoué (1989) and Inoué et al. (1997),

---

*In endocrinology, the term "hormone" is often reserved for a substance that has been chemically identified and has met the criteria established for being a hormone. The term "chemical factor" is used for a substance that has not been chemically characterized and has not qualified as a hormone.

Ishimori, in 1909, deprived dogs of total sleep for up to 113 hours. Their brains were removed, processed, and desiccated, with the dry substance then rehydrated and injected subcutaneously into recipient dogs. The dogs exhibited a number of effects, one of which was excessive sleep. Control dogs that received brain extracts from non-sleep-deprived animals showed some of the same effects, but not increased sleep (Ishimori, 1909).

Between 1907 and 1912, Henri Piéron and his colleague R. Legendre, in Paris, generated a series of papers that was summarized and elaborated upon in a book (Piéron, 1913). Piéron coined the term "hypnotoxin" to refer to a chemical that would accumulate in the brain during waking and cause drowsiness and sleep. More neutral terms for such a sleep-inducing agent are "hypnogenic" and "somnogenic." Piéron and Legendre, according to Jouvet (1972), deprived dogs of sleep by walking them through the streets of Paris, after which the experimenters removed a small amount of CSF and injected it into a cerebral ventricle of non-sleep-deprived dogs. The recipient animals promptly fell asleep. Control animals, receiving CSF from non-sleep-deprived dogs, did not fall asleep.

In the 1940s and 1950s, Marcel Monnier and his colleagues in Basel, Switzerland, described related findings. Monnier et al. (1963) performed crossed circulation experiments, in which the blood from one rabbit was delivered to the circulation of a recipient rabbit. When the intralaminar thalamus of the donor animal was stimulated with parameters that produced EEG slow waves, the recipient rabbit also showed increased slow waves. Since the circulatory system was the only link between the two rabbits, the clear implication was that the synchronogenic stimulation in the donor rabbit caused the release of a blood-borne chemical factor that, in turn, caused the synchrony and sleep in the recipient animal. By extension, they concluded that the sleep-induced in the donor animal by electrical stimulation was mediated by that same chemical factor.

In the 1960s and 1970s, Pappenheimer and colleagues at Harvard University collected CSF from the cisterna magna of sleep-deprived goats. Goats were used because they provided an abundant supply of CSF under nonstressful and nonanesthetized conditions. The collected CSF was infused into the ventricles of cats and rats. The recipients showed the behavioral signs of normal sleep; that is, they could be awakened by sensory stimuli, but fell asleep again. As in the work of Piéron, control injections from non-sleep-deprived animals had no effect. Pappenheimer commented that "the fact that fluid from sleep-deprived goats is active in cats and rats suggests that we are dealing with a humoral factor of general and fundamental importance to the sleep mechanism" (Pappenheimer, 1967, p. 516). Experiments of these types have since been repeated and in many cases the results have been replicated, but in some cases they have not been. The reason for this discrepancy is not clear. However, the positive findings are sufficiently persuasive to indicate that the basic idea of a chemical sleep factor is sound. The challenge is to determine what it is (if there is only one), where it is synthesized, and where in the brain it has its effect.

The technique of dialysis has been used to isolate the active compound in blood and CSF. Dialysis is a procedure in which a fluid (blood, CSF, or urine) is filtered and particular components are extracted. Schoenberger and Monnier (1977) reported that a short peptide isolated from the blood of rabbits in slow-wave sleep was effective in producing sleep in recipient rabbits. They labeled this compound "delta-sleep-inducing peptide" (DSIP). Jouvet and his colleagues in Lyon had similar effects with a chemical they called "sleep-promoting substance" (SPS) extracted from the brains of sleep-deprived rats and injected systemically into mice (Nagasaki et al., 1980).

At least two observations complicate the issue of a sleep-inducing factor. One is that Siamese twins, who share the same blood circulation, do not always sleep at the same time. One twin may become tired and sleep while the other remains awake and alert. Second, when a person is forced to stay awake all night, the urge to sleep is most intense between about 2 A.M. and 6 A.M. However, in the morning when waking usually occurs, the individual brightens up for a number of hours and the drive to sleep is much reduced. In both cases if a sleep chemical is present, other variables must also exert an influence. For the common circulation of Siamese twins it may be, as pointed out by Kleitman (1963, p. 354), that an aroused psychological state of the brain of one twin could serve as a counterinfluence to the effects of a blood-borne factor so that sleep would not occur as readily. For the brightening effect in the morning when one normally awakens, it is clear that the imposition of a circadian rhythm is at work. The biochemical changes associated with this phase of the circadian rhythm appear to supersede the effect of sleep deprivation—at least for a period of time. However, as the day wears on, the circadian rhythm shifts toward the production of physiological changes associated with sleep. The sleep-inducing factor accumulated during sleep deprivation finds a brain conditioned, so to speak, by these circadian changes, and the urge to sleep becomes progressively stronger.

As stated earlier, workers in the field would like to know where one or more sleep-inducing factors are synthesized and what pathways lead from those sites to the locations in the brain at which these chemicals have their effects. Given the great complexity of the circuitry of the brain, a current approach sidesteps the issue of where the chemical agent (neurotransmitter or hormone) is produced and how it gets to where it has its effect. This approach concentrates on the site of action of the agent. For a physiological effect to occur, the neuroactive agent must bind to a receptor. This point can be made most readily for hormones. Hormones are released from endocrine glands and enter the circulatory system as components in blood. As such, the hormones are ubiquitous throughout the body but will affect only cells that have receptors to which the specific chemicals have an affinity. In this context, the chemical may be referred to as a ligand. If the hormone affects nerve cells, we may call it a neurohormone. In the nervous system there are different receptor subtypes to which the same ligand (hormone or neurotransmitter) will bind, and in each case the effect will be specific to the receptor subtype. However, there will be an effect only where there are receptors for the chemical. Advances in biochemistry and molecular biology have been remarkably successful in identifying a multitude of receptors, receptor subtypes, ion channels, and endogenous ligands that bind to nerve cell receptors. To complement these developments, and in some cases as instruments of the developments, pharmacologists have created agonists and antagonists for these myriad receptor subtypes that have permitted experimental manipulations of these hormone– and transmitter–receptor-systems.

Sleep scientists are capitalizing on these advances by depositing minute amounts of receptor agonists and antagonists into specific nuclei of the brain to produce alterations of sleep and waking. Emerging from this work, which has obvious relevance to our discussion of sleep factors, are findings that areas of the brain that have long been implicated in slow-wave sleep on the basis of electrical stimulation can be influenced to produce sleep by the application of a variety of neuroactive agents. Some are conventional neurotransmitters, like acetylcholine, whereas others are novel substances. We should keep in mind that electrical stimulation of the brain, albeit a very effective way of activating neural tissue, is an artificial way of doing so. The physiological or normal means by which brain cells are affected is, of course, by neurotransmitters and hormones—not by thunderbolts from afar (as those critical of electrical stimulation would describe it). As examples of experimental approaches that are more physiological, I will describe the work of James Krueger and Robert McCarley, two sleep scientists who recently provided important information on the biochemical control of sleep.

James Krueger, at Washington State University, worked with Pappenheimer at Harvard. His early papers with the Pappenheimer group were the first in a line of research in which Krueger developed a fascinating story that related the immune system to sleep. Recall that Pappenheimer and others had demonstrated that a blood-borne factor from sleep-deprived animals could promote sleep. Krueger et al. (1985) related this to the observation that people and animals who become sick from bacterial infection show increased sleep and increased body temperature. Krueger and his coworkers extracted and identified the active agent in urine from bacterial-infected human subjects that, when injected into rabbits, produced increased amounts of sleep. The substances in question were muramyl peptides, which are protein residues of digested bacteria, the skeletal residues of the cell walls of killed bacteria. When phagocytic cells, the first line of defense of the immune system, incorporate and digest disease-causing bacteria, the muramyl peptides are released into the circulatory system. These peptides have the effects of stimulating an immune response and producing an elevated temperature (fever) and sleep.

Since the focus of Krueger's work was sleep, it was of interest to know whether sleep was secondary to the fever, that is, whether sleep was caused by the elevated temperature. This is a common interpretation of sleep accompanied by fever. The simple manipulation of administering an antipyretic (a fever-reducing drug, like aspirin) answered this question. Fever was eliminated, but excessive sleep still occurred with the microbial infection.

The next step was to discover what mediated the connection between microbial products and the sleep response. In the early 1990s it was discovered that muramyl peptides stimulate the immune system to synthesize and release a number of cytokines, protein molecules that produce sleep and fever. The cytokines that had this effect were shown to be interleukin 1 (IL-1), interferon alfa (IFN-α), and tumor necrosis factor (TNF). These cytokines stimulate the hypothalamus to produce the growth hormone releasing hormone (GHRH), which in turn stimulates the secretion of the growth hormone itself from the pituitary gland. However, most interestingly, the GHRH has also been implicated in sleep. Krueger and colleagues have recently shown that GHRH injected into the basal forebrain–preoptic area of rats increases slow-wave sleep (Zhang et al., 1999). Its normal sources in the brain are two groups of cells in the hypothalamus. One group of cells provides the releasing hormone to the anterior pituitary via the hypophyseal portal system for the subsequent release of the growth hormone. The other population of cells is in the ventromedial nucleus of the hypothalamus, which projects to the basal forebrain–preoptic area. The preoptic area has been known to be a region responsible for sleep since the time of von Economo. In this case it appears that GHRH should be characterized as a neurotransmitter or neuromodulator, since it is released from axon terminals and has a synaptic effect on nerve cells of the basal forebrain–preoptic area. Finally, completing this line of reasoning, the extended basal forebrain area, with its GABAergic population, exerts an inhibitory drive upon histamine cells of the posterior hypothalamus and cholinergic cells of the pontomesencephalic reticular formation, two major components of the arousal system. The end result would thus be sleep.

Some insight into the utility of this system may be derived from Krueger's observations on rabbits that received a bacterial challenge. Some rabbits got sick, deteriorated, and died. Others got sick but recovered. Those that died showed only a short period of sleep enhancement and deteriorated rapidly. In contrast, the survivors displayed longer periods of sleep enhancement. The inference drawn was that a rabbit with a more competent immune system could combat the microbial challenge by the greater synthesis of cytokines and survive. Rabbits treated with immunosuppressive drugs showed less sleep in response to the infection and a lowered rate of survival. However, it is still not known whether sleep per se has the salutary effect.

Another approach to the role of chemical factors in sleep is one taken by Robert McCarley and his Harvard colleagues, who have explored the sleep effects of adenosine, a purine product derived by the breakdown of adenosine triphosphate (ATP) during metabolic processes and implicated in sleep since the 1980s. More recently, McCarley and coworkers reported in the journal *Science* (Porkka-Heiskanen et al., 1997) that extracellular fluid samples collected from two regions of the cat brain during sleep and waking showed increased amounts of adenosine during waking. This effect was greater after prolonged wakefulness. During sleep recovery, adenosine concentrations decreased. This finding is not surprising because, as the authors point out, adenosine is a breakdown product of cellular metabolism and it is known that brain cell activity and thus brain metabolism are greater during waking than in slow-wave sleep (Maquet et al., 1990).

The two areas from which samples were collected were the basal forebrain, known to be involved in sleep/waking control, and the ventral anterior–ventral lateral (VA-VL) complex of the thalamus, which has not been implicated in sleep. McCarley and his colleagues also experimentally manipulated the extracellular concentrations of adenosine in these two brain regions. Only increased adenosine in the basal forebrain induced an increase in slow-wave sleep; thalamic VA-VL increases were ineffective. What makes the thalamic lack of effect interesting as more than a control area for the experimental increase of adenosine in the basal forebrain, is that adenosine is one of the most ubiquitous substances in the brain. As a breakdown product of ATP during ATP's role as the major source of energy for cellular activity, adenosine is present everywhere, not only intra- but also extracellularly, and it shows an increase during waking in the VA-VL area as well as in the basal forebrain. It appears that an important difference between these two brain areas, relevant to sleep, is that cells of the basal forebrain, in contrast to those of VA-VL, possess adenosine receptors to which adenosine will bind. One class of adenosine receptor, the A_1 receptor (also referred to as the P_1 or purinoreceptor), controls a potassium channel that permits an influx of the negative potassium ion and results in a hyperpolarization (inhibition) of the cell. Since it is known that increased activity of the cholinergic basal forebrain produces arousal, the notion is that during sustained wakefulness and the increased metabolic activity of these arousal system cells, the adenosine generated may serve as a sleep factor by feeding back upon receptors of these same cells to inhibit their activity and thus reduce wakefulness. A direct test of this was conducted by McCarley's group (Thakkar et al., 1999) by recording from arousal system cells of the basal forebrain. The investigators applied different concentrations of an A_1 receptor agonist locally to those cells and found a dose-dependent decrease in their firing rate. This demonstrated that adenosine does in fact inhibit the firing of arousal-producing cells of the basal forebrain. There is evidence also of adenosine receptors on cholinergic cells of the brain stem mesopontine arousal area that function in a similar, negative feedback fashion.

In summary, an emerging view is that multiple brain areas are involved in the control of sleep, and multiple chemical factors in the form of neurotransmitters, neuromodulators, and blood-borne neurohormones control the neural activity of these brain areas. All effects, however, are mediated by receptors on specific populations of neurons, which in turn drive those cells into a state of activity that produces sleep. In all cases, it appears that the final step in producing sleep is the inhibition of cells responsible for arousal.

10

Sleep as a Circadian Rhythm

A number of behaviors and physiological functions occur in rhythmic cycles of different durations. For example, biological rhythms include the long seasonal breeding rhythms and shorter estrous and monthly menstrual cycles, during which hormonal and reproductive activities occur. And, relevant to our topic at hand, there are a number of 24-hour cycles, one of which is an activity cycle that includes the sleep–waking rhythm. This daily rhythmic cycle is not precisely 24 hours; rather, its length is about (*circa*) a day (*dies*), hence the term "circadian rhythm."

It is both surprising and interesting that research in circadian rhythms and in sleep/waking developed independently. Until recently, researchers in these two areas have maintained separate professional societies, and few in each discipline attended the meetings of the other. Fortunately, this is changing. In 1999 a conference was organized with the explicit goal of bringing workers in the two areas together to discuss the interactions between the disciplines. A focus of the conference was the anatomical and physiological means by which sleep/waking and circadian rhythms are coupled.

The classic illustration of a circadian rhythm is the activity level of a rat represented on a 24-hour time line. Rats are nocturnal animals: they eat, drink, and make merry when it is dark and are relatively inactive and sleep during the light phase of the 24-hour period. As depicted in Figure 10.1, if activity is indicated by small vertical tick marks on a 24-hour horizontal time line, dense clusters of ticks occur during the dark period and a contrasting paucity of tick marks is seen during the light period when the rats are inactive and asleep. As Figure 10.1 shows, when successive 24-hour time lines are stacked one below the

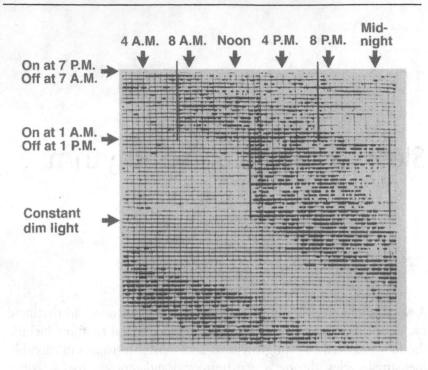

FIGURE 10.1. Activity cycles of a rat entrained to light and dark periods and during constant illumination. Vertical lines have been added to clearly show the times of light onset and light offset. (Adapted with permission from a paper presented at a meeting of the Eastern Psychological Association, 1980, by T. A. Groblewski, A. Nuñez, and R. M. Gold.)

other, and the room lights are switched off at 7:00 A.M. and on at 7:00 P.M., a clear pattern of activity levels reveals itself. During the duration of the dark period there is a high level of activity; when the lights are turned on at 7:00 P.M. there is a sudden decrease of activity, which lasts throughout the 12-hour light phase. This pattern is very discernible when the 24-hour days are stacked successively.

When the period of illumination is shifted to lights-on at 1:00 A.M. and lights-off at 1:00 P.M., after a few days the activity cycle settled down to a new periodicity. It is clear that illumination is governing or entraining the activity cycle. What happens if the animals are kept in constant darkness or constant dim illumination? Is the cyclical activity level controlled by the external occurrence of light/dark changes, and in the absence of such changes does the activity occur intermittently and randomly throughout the 24-hour period? Interestingly, the circadian rhythm stays intact, with a periodicity close to 24 hours. In the case of the illustration of Figure 10.1, the animal showed a free-running cycle of approximately 25 hours. In the absence of an external timekeeper to entrain the cycle (i.e., in continuous dim light), this animal's intrinsic rhythm gained about one hour per day. In the display, the activity cycle drifted to the right by that amount each day. These data tell us that there is a near-24-hour activity rhythm that is genetically programmed and that it can be entrained by external events, in this case changes in illumination level.

An intrinsic rhythm 24 hours in duration suggests that this pattern is governed by a 24-hour biological clock. A major advance in the field was the discovery of the location of the biological clock. In 1972 Robert Moore and Irving Zucker and their colleagues independently demonstrated that the suprachiasmatic nucleus (SCN) of the hypothalamus was the biological clock (Moore and Eichler, 1972; Stephan and Zucker, 1972). Bilateral lesions of this nucleus in the rat totally abolished the circadian sleep–waking cycle. Light and dark periods no longer entrained the activity levels normally seen during sleep and waking. Figure 10.2A shows the region of the SCN, and Figure 10.2B plots the activity levels of a rat before and after lesions of the SCN.

Since light is the primary controlling stimulus for the biological clock's daily rhythm, one would expect that the clock receives an input from the visual system. It turns out that there are two visual inputs to the SCN. The primary one is directly from the retina. The fibers of the optic nerve from each eye meet at the base of the brain just below the hypothalamus. This is the optic chiasm, where some fibers cross and project to the contralateral thalamus and others continue on to the ipsilateral thalamus. The dorsal portion of the lateral geniculate nucleus of the thalamus receives visual information from the retina and relays it to the visual cortex. A select few of the fibers from the retina leave the visual pathway at the optic chiasm and turn upward into the SCN of the hypothalamus. This is the retinohypothalamic tract. If the optic tract between the chiasm and the thalamus is severed, the animal is rendered blind but still shows circadian rhythms entrained to light. If the retinohypothalamic tract is severed, light entrainment is lost, but vision is unimpaired.

A second input to the SCN is from the intergeniculate leaflet of the ventral part of the lateral geniculate nucleus of the thalamus. This portion of the lateral geniculate does not relay retinal information to the cortex, as does the dorsal lateral geniculate. The retinal input to the SCN via the intergeniculate leaflet appears to serve a modulatory rather than a primary control function on circadian rhythms.

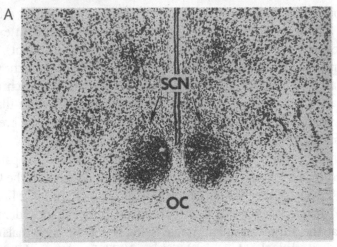

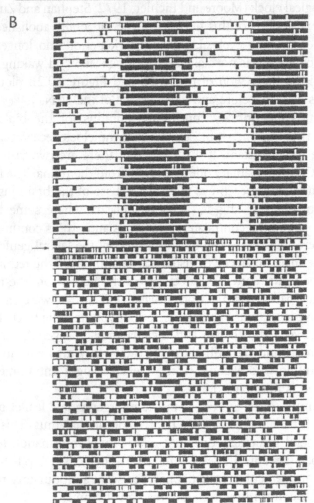

Thus the SCN is necessary for circadian rhythm/entrainment to light, but we saw earlier that even without light or with a constant illumination level, circadian rhythms still occur—albeit as a free-running rhythm close to 24 hours in length. Except for species that evolved and live in the dark depths of the sea or in deep dark recesses of the land, as in caves, activity in animals is governed by light-entrained cycles. The SCN, even without entraining stimuli, maintains cyclical functions. How do the cells of the SCN accomplish that? There are a number of ways that a population of cells can fire together synchronously and in a cyclical pattern. They can be synaptically connected in simple to complex circuits with appropriate positive and negative feedback and feed-forward loops. Alternatively, cells may have the intrinsic property of pacemaker cells with a specified firing rhythm. We described such a cellular mechanism in earlier chapters in the context of the thalamic reticular nucleus and thalamocortical synchronization. It turns out that both mechanism (circuits and pacemakers) apply, and each serves an important part of the functioning of the SCN.

Two salient features of pacemaker cells are rhythmicity and synchronicity. These cells exhibit an increased level of activity for a period of time followed by reduced or no activity—and the cycle repeats. The repeated on–off pattern is the rhythmicity feature of the cycle. For such cells to form an effective influence, they must also fire synchronously with each other. The rhythmicity property of pacemaker cells is intrinsic to the cells. They need not be externally driven, for they are inherently programmed to fire that way. Recent work has described the expression of particular genes for proteins that appear to govern this property. The firing of cells of the SCN in their rhythmic fashion in unison (synchronously) is not an intrinsic property of the cells. The intrinsic and nonintrinsic properties of the SCN pacemaker cells have been demonstrated by the following experiments.

FIGURE 10.2. (A) Frontal section through the hypothalamus of a rat showing the location of the suprachiasmatic nucleus (SCN), a bilateral structure comprising dense clusters of cells directly above the optic chiasm (OC). The slitlike structure between the left and right hypothalamus is the third ventricle. (Reproduced with permission from Moore, 1999, p. 1192.) (B) Activity record of a rat before and after lesions of the SCN. Each horizontal time line represents two 24-hour periods, with each 24 hours having 12 hours of light followed by 12 hours of dark. The top half shows activity levels that are clearly entrained and synchronized by the presence and absence of light. High activity levels are indicated by the dark areas. The bottom half shows activity after lesions of the SCN. It is equally clear that activity and rest periods are randomly distributed throughout the 24-hour periods; there is no longer a circadian rhythm. (Reproduced with permission from Moore, 1999, p. 1191.)

If a brain slice containing the SCN is maintained in a bath with oxygen and provided with nutrients, nerve cells can be kept viable and neural activity can be recorded from them. In such a brain slice preparation, cells of the SCN behave as they do in the intact brain. They fire rhythmically and in unison. This tells us that external influences are not necessary for these two properties to manifest themselves.

Another preparation proved very revealing. Cells of the SCN were placed in a culture medium and dissociated from each other by adding an enzyme that digests tissues that maintain connections between cells. Neurons of the SCN were thus isolated from each other. Recordings taken simultaneously from two or more cells showed that each cell still fired in a rhythmic fashion, but the firing patterns of the individual cells were independent of each other. They did not fire synchronously (Welsh et al. 1995). Cell interaction is necessary for this property.

An anatomical feature of SCN cells provides a possible mechanism for the interaction just described. Cells in the part of the SCN that are most responsible for the cyclical firing pattern have profuse dendritic arbors that intermingle in a dendritic matrix. These dendrites are endowed with dendrodendritic synapses. This unusual form of cell coupling may be the mechanism that accounts for the synchronous firing of the SCN neurons.

The final and, of course, important question is how the SCN exerts its timekeeping influence on the parts of the brain that control sleep and waking. The direct efferent projections of the SCN are limited to diencephalic and telencephalic structures. The heaviest projection is to a nearby region of the hypothalamus dorsal to the SCN, the paraventricular nucleus and an area just ventral to it, called the subparaventricular zone. Since cells of the subparaventricular zone and the SCN project essentially to the same targets, it appears that a function of the subparaventricular zone is to reinforce the action of the SCN.

There is nothing subtle or indirect about the connections between the SCN and areas that control sleep and waking. The SCN and the subparaventricular zone project directly to the cholinergic basal forebrain and histaminergic posterior hypothalamic systems that subserve cortical arousal. The SCN and subparaventricular zone also project to the preoptic/anterior hypothalamic region that mediates sleep via its GABA-inhibitory influence on the posterior hypothalamic arousal system. The nature of the neural information conveyed by the SCN fibers is also relatively straightforward. During daylight the frequency of the spike-firing rate is about twice as great as it is during the night.

Strangely, this firing pattern is the same in nocturnal and diurnal animals. It is not known how the same frequency-coded messages are translated at their target cells to produce their opposite effects.

Another mystery enters the story at this point. We recall that circadian rhythm functions are lost if the SCN is lesioned. However, if SCN cells from another animal are transplanted into the lesioned animal, circadian functions are restored. This is true even if the SCN implant is not at its normal location and does not reestablish its usual neural connections. This was shown most convincingly by Rae Silver and her colleagues (1996) with an SCN implant contained in a semipermeable capsule that was inserted into the third ventricle. This device permitted SCN cell survival and humoral communication across the membrane, but neural connections could not be established. Apparently, the SCN has hormonelike properties that permit chemical communication to appropriate neural targets to orchestrate the circadian responses that normally occur by neural connections. LeSauter and Silver (1999) later showed that a specific subnucleus of SCN is critically important for circadian rhythmicity. They reported that "transplants . . . that contain cells (of this nucleus) restore locomotor rhythms in SCN-lesioned host animals, whereas transplants containing SCN tissue but lacking cells of this subnucleus fail to restore rhythmicity" (LeSauter and Silver, 1999, p. 5574).

In addition to the neural and humoral influences described, the SCN affects the sleep–waking cycle by a connection to the pineal gland. This connection is mediated by the paraventricular nucleus, which has a long projection from the hypothalamus through the entire brain stem to the thoracic level of the spinal cord. These axons synapse with a population of cells in the intermediate horn of the spinal cord (intermediate between the dorsal and ventral horns) that project fibers out of the cord to the superior cervical ganglion. This ganglion lies alongside the spinal cord at the cervical level and is the uppermost of the sympathetic chain ganglia. Some postganglionic cells originating in the ganglion project fibers into the brain to terminate in the pineal gland. The pineal body emanates from the dorsal and posterior part of the thalamus via a midline stalk, and the gland eventually rests at the rostral mesencephalic level above the tectum. It is somewhat unique in the nervous system in that it is a single midline structure.

Under the influence of the sympathetic noradrenergic input, the pineal gland synthesizes and releases the hormone melatonin. At night, during the dark, the pineal gland is maximally active to secrete melatonin. During the day when the SCN receives a retinal input from light, the neural message it relays to the pineal via the sympathetic system results in a cessation of melatonin production and secretion.

A major function of the pineal gland and melatonin is the control of the breeding cycle. In sheep and some other species that have a long gestation period, breeding starts with mating behavior that occurs in the fall and winter when the days are short. In other species, like the hamster, which have a short gestation period, the breeding cycle occurs in the spring and summer when days are long. The utility of the differences between the long and short breeding cycles with respect to the time of mating and the duration of the gestation period is that in both cases parturition will occur during warm weather, which is optimal for survival of the newborn. In both breeding cycles the critical stimulus that triggers the cycle is the amount of daylight. In short-day breeders, the short periods of daylight are sensed by the SCN and the information is relayed to the pineal gland. During short days and long nights, melatonin secretion will be high, since the synthesis and secretion of melatonin occurs during dark periods. The hormone will affect the hypothalamic nuclei that have neurons with melatonin receptors. One of these, the infundibular nucleus, activates the hypothalamic–pituitary connection, which in turn triggers the release of anterior pituitary gonadal hormones that initiate the reproductive cycle.

In the case of long-day breeders, like the hamster, short nights result in low levels of melatonin. In these animals, a low melatonin level is the critical condition for activation of the hypothalamic–pituitary axis governing reproductive activity. How the different amounts of circulating melatonin produce opposite effects in different species is not known.

The pineal gland and melatonin influence not only seasonal cycles, but also daily cycles that include the sleep–waking cycle. Melatonin receptors are located not only on hypothalamic neurons governing the pituitary, but also on cells of the SCN. The light sensory nucleus that starts and governs the process of melatonin secretion is also affected by melatonin.

Melatonin is one of a number of influences besides light that control the neural activity of the SCN, the synchronizer of circadian rhythms. Melatonin, administered at appropriate times in the circadian cycle, will resynchronize the cycle to another period. How melatonin works to alter circadian rhythms is still debated by researchers in the field. Nevertheless a practical application of this finding has been used as a "treatment" for jet lag. The procedure is tricky because the times and amounts of melatonin administration depend on the direction of travel (east vs west) and the number of time zones traversed. In addition, the times and amounts of melatonin administered are different on the days preceding and following the trip. Currently, there is no agreed-upon formula, so any announcement of a melatonin treatment for jet lag is premature.

The phenomenon of jet lag emphasizes that the sleep/wake cycle is controlled by two components (or processes), a homeostatic and a circadian one. Alexander Borbély and colleagues, in Zurich, Switzerland, have formalized and detailed the nature of the interaction between the two components (Borbély, 1982: Borbély and Achermann, 2000). The homeostatic process refers to the increasing need for sleep the longer one stays awake, incurring a sleep debt. Similarly, the need for sleep decreases the longer one sleeps. The circadian component refers to the intrinsic 24-hour biological clock that evolved in humans and other animals on our planet Earth, which has a 24-hour day/night cycle. As diurnal animals, we have a circadian clock that is set to awaken us at the onset of the light period and prompts us to sleep at night. The homeostatic and circadian components are independent of each other; normally, however, they are "in synch" and we are not aware of the separate influence of each. It becomes apparent that these components are separate when we experience jet lag—that is, when the two components are no longer "in synch." It takes a few days in the new time zone for the homeostatic and circadian components to catch up to each other and become synchronized again.

Part II
The Second
Half-Century: The
Benefits Are Reaped

Section 3
The Dreaming Brain:
REM and Paradoxical Sleep

11

The Discovery of REM Sleep

Nathaniel Kleitman (1895–1999) may be considered to be the grand-father of sleep research. During his long career at the University of Chicago, he generated important research papers on multiple aspects of sleep, a major book on sleep and, importantly, students who became leading researchers in the field. Kleitman's book, *Sleep and Wakefulness*, was published in 1939. By the late 1950s, the book had been out of print for a few years. Sparked to a large extent by research reports from Kleitman's own lab in the 1950s, there was renewed interest and demand for his book, and in 1963 he published a revised and enlarged edition. This was a tome of 552 pages with a bibliography of 4337 entries! For an exhaustive and scholarly review of the sleep and waking literature up to that date, there is no equal. To the extent that Kleitman's name is familiar to those who aren't sleep researchers, it is due to his coauthorship with Eugene Aserinsky of the paper published in 1953 in the journal *Science* that described their discovery of rapid eye movement (REM) sleep (Aserinsky and Kleitman, 1953). The account of this important discovery has been gleaned from the paper itself and a number of other sources, including Kleitman's retrospective description of it in his 1963 book and Aserinsky's recollections published more recently (Kleitman, 1963; Aserinsky, 1996).

In the early 1950s Eugene Aserinsky (1921–1998) was a graduate student in Kleitman's lab studying the relationship between slow eye movements and the depth of sleep. In infants he was able to detect these eye movements by direct observation, since infants sleep very readily in a lighted room and his presence did not interfere with their sleep. Aserinsky noted that the slow eye movements at the beginning of sleep soon gave way to rapid eye movements that coincided with reduced body movements. When sleep and associated phenomena were monitored in adults, it became important for the room to be dark and observations to be indirect so as not to disturb the sleeper. Aserinsky and Kleitman succeeded in monitoring eye movements remotely by attaching a pair of small electrodes to the skin medial and lateral to one eye and another electrode pair below and above the eye. Differences in electrical potential between the components of an electrode pair were generated whenever there were horizontal or vertical eye movements, respectively. A recording of this type is an electro-oculogram (EOG). Electrodes were also attached to the scalp for recording the EEG. Figure 11.1 shows Nathaniel Kleitman as a subject in one of his EOG-EEG sleep studies. Wires leading from the electrodes were inserted into an electrode board fixed to the headboard of the bed. After an adaptation period, the wires, which were very light and flexible, did not interfere with normal sleep. The wires from the electrode board carrying the very low voltage signals were connected via an electrically shielded cable to amplifiers of the EEG machine located in an adjacent room.

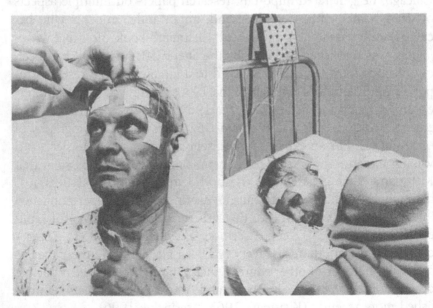

A   B

FIGURE 11.1. Photographs of Nathaniel Kleitman as a subject in a sleep study. (A) Electrodes being attached to Kleitman's scalp for recording the EEG and electrodes attached to his face just lateral to and above each eye are for the recording of eye movements. (B) Kleitman sleeping with the electrode leads inserted into an electrode board, which in turn carries shielded wires into an EEG machine in an adjoining room. (Adapted from N. Kleitman, Patterns of dreaming, Scientific American, November 1960, p. 83.)

Noise from the EEG machine with its moving paper and scratchy pens, and the experimenters who monitored the arrangements, would have disturbed a person trying to sleep in the same room.

The EOG electrodes periodically registered large, rapid potentials indicating vertical or horizontal eye movements. To be sure that these large voltages were not artifacts, Aserinsky directly observed some of the subjects under dim illumination to determine whether, in fact, there were eye movements when the EOG tracings showed these deflections. It was verified that rapid eye movements occurred during sleep. When the EEG was also recorded, Aserinsky discovered that during sleep when there were periods of eye movements, the EEG showed a waking pattern: that is, a low voltage, rhythm of irregular frequency, similar to beta waves. The suspicion was that this was a period of dreaming. As stated in the original 1953 paper, "To confirm the conjecture that this particular eye activity was associated with dreaming, . . . (subjects) were awakened and interrogated during the occurrence of this eye motility and also after a period of . . . ocular quiescence." On the basis of their observations, Aserinsky and Kleitman concluded that "the ability to recall dreams is significantly associated with the presence of the eye movements noted" (Aserinsky and Kleitman, 1953, p. 274).

It is interesting at this point to note that the human EEG with its basic sleep and waking rhythms was first reported by Berger in 1925 and for almost 30 years (up until 1953), it was assumed that there were brief periods of awakenings when the EEG in the adjoining room to the sleeping person showed periods of low voltage, fast activity (LVFA). This assumption made logical sense: during sleep recordings the periods of awakenings were brief during the beginning hours of deep delta sleep, and as the night wore on and sleep/rest time accumulated, sleep lightened and the durations of EEG arousal periods increased. Apparently, Aserinsky was the first to look in on sleeping subjects and discover that during these periods of LVFA, the person, in fact, had not awakened but was still asleep! Aserinsky and Kleitman suggested that dreaming occurred during these periods. (This story emphasizes that there is no substitute for careful observation in experimental research.)

About this time, William Dement, a medical student at the University of Chicago, joined Kleitman's lab. His first project was to awaken people during various periods of sleep and determine whether if they had been dreaming. In 1957 Dement and Kleitman published two papers in which they confirmed and detailed the relationship between rapid eye movements and dreaming (Dement and Kleitman, 1957a,b).

During this exciting period of sleep research, one of the interests was to determine how general a phenomenon dreaming was. This was first studied as part of a larger project by Dement (1955a,b) in which subjects were asked whether they dreamed. A small number reported that they did not, or did so infrequently. When these subjects were submitted to all-night sleep recordings without being disturbed, their polygraph records showed the usual four or five REM episodes that the other subjects showed. However, in the morning the alleged nondreamers reported that, as usual, they had not dreamed. When, on subsequent nights these subjects were awakened during a REM episode, they reported that they had experienced a dream! Apparently, there are no nondreamers; but there are non-dream-recallers. This and subsequent studies established that REM sleep occurs in all humans studied and has a typical periodicity during the night (refer to the hypnogram in Chapter 2, Figure 2.2). The first REM episode occurs about an hour to an hour and a half after falling asleep and lasts a few minutes. Subsequently, episodes occur at about 90-minute intervals throughout the night and, as the night progresses, the episodes become longer. The last REM period, prior to awakening in the morning, may be as long as 30 to 50 minutes.

Some of the findings and conclusions of the early work on the association between REM sleep and dreaming are being reevaluated. In the early research papers there were often comments to the effect that not every awakening during a REM episode produced a dream report. In fact, between 70 and 80%, not 100%, of the awakenings from REM sleep were accompanied by dream reports. Also, about 10% of the awakenings from non-REM sleep carried dream reports. It was assumed that some dream content was relatively inaccessible to recall and that some subjects were poor dream recallers. The relationship between REM sleep and dreaming was too comfortable to give up. The consensus was that dream sleep and REM sleep were essentially two aspects of the same brain state.

Recent work, however, indicates that the relationship is not as robust as was previously thought. Reports of dreams upon awakening from non-REM sleep are as high as 70%. Not all these dream reports can be dismissed as recollections of dreams that occurred during a previous REM episode because some of the reports were from the start of sleep before the first REM period. There are some differences between REM and non-REM dream reports, but they do not appear to be major differences. The daunting task for the sleep researcher, yet to be developed, is to determine which of the neural events that occur during REM sleep are also part of the neural substrate essential for REM sleep dreaming and whether these, or other neural events, are necessary for non-REM dreams.

The publications in the 1950s from Kleitman's laboratory, as one would imagine, opened a floodgate of interest and experiments. One area of interest was the function of dream sleep. If we all dream, dreaming must have some utility. Dement, whose interest as a medical student was psychiatry with a Freudian bias, theorized that as we internalize parental and societal values and thus develop a superego, many of our desires and urges are not consciously acknowledged, but repressed. To avoid the neurotic outcome of wishes and desires repressed because we have learned that they are deemed socially unacceptable, the dream process serves to give expression to them, albeit in often disguised and symbolic form. Dreams, therefore, prevent us from becoming neurotic—if not worse. Related to this, Dement speculated that the bizarre thought and perceptual processes experienced by schizophrenic patients may be due to a failure to dream during sleep, and the subsequent pressure to dream manifests itself during waking as psychotic episodes. Imagine someone observing you acting out a dream episode while you're awake. You would be judged as having a psychotic episode. Dement's creative idea became the basis of a graduate research project at the University of Chicago (Dement, 1955b). He did all-night sleep recordings on schizophrenic patients to see if they had REM sleep. He found that the sleep of schizophrenics, including REM sleep, was barely different from that of normal subjects.

Dement was undaunted in his search for the function of REM sleep. When his medical and Ph.D. training were completed in Chicago, he took a position in psychiatry at the Mount Sinai Hospital in New York City. There he did the critical "lesion" experiment. In neurobiology, one of the strategies for discovering the function of a part of the brain is to surgically remove it and see how behavior is different without that brain area present. The equivalent experiment for Dement was to remove REM sleep by preventing subjects from having it (Dement, 1960). Eight volunteer subjects were acclimated to all-night sleep recordings and were told that they would be awakened periodically during the night for five consecutive nights. They were not told when or why the awakenings would occur. For all subjects, Dement did the REM sleep deprivation part of the experiment first to determine the number and pattern of awakenings required for each subject to prevent REM sleep. This information was used later for control awakenings from non-REM sleep. Subjects were awakened as soon as Dement detected the onset of a REM sleep episode. Of the eight subjects, one quit the experiment in an agitated state after three nights of REM sleep deprivation. Two subjects would not continue after four nights but agreed to continue with recovery nights, during which they were permitted to sleep without being awakened. Four subjects completed the five nights of REM sleep deprivation and one subject agreed to continue for seven nights.

As a psychiatrist, Dement was a trained observer of behavior and relied on his interactions with the subjects to detect alterations from their normal behaviors. He reported that REM sleep deprivation produced anxiety, irritability, and difficulty in concentrating. In subsequent accounts of this and additional REM sleep deprivation experiments, Dement indicated that some subjects showed signs of psychological imbalance, increased appetite and activation level, and an increased interest in normally taboo sexual subjects (Dement and Fisher, 1963; Fisher and Dement, 1963; Dement, 1965). On control nights when subjects were awakened from non-REM sleep with as close to the same number and pattern of awakenings required to prevent them from having REM sleep, the psychological disturbances were not seen. These findings supported some of Dement's original ideas at Chicago. However, the psychological effects of REM sleep deprivation were not consistently replicated by subsequent researchers or by Dement himself. To this day, it is not clear why some subjects appeared to be affected by REM sleep deprivation and others not. The consensus among sleep workers, based now on many such studies, is that a number of nights of REM sleep deprivation does not cause psychological disturbance.

Another finding by Dement reported in the 1960 *Science* paper turned out to be very robust and of great importance, and has been replicated many times. On each successive night of REM sleep deprivation, subjects had to be awakened more frequently to prevent them from entering REM sleep, and the number of REM sleep attempts increased on successive nights of REM sleep deprivation. Also, after the first night of REM sleep deprivation, the first REM episode appeared before the end of the usual 90-minute period, and with more nights of deprivation, the first REM sleep attempt occurred earlier and earlier. When the nights of REM sleep deprivation were over, subjects were permitted to sleep without any imposed disturbances. On these recovery nights, subjects went into REM sleep shortly after the beginning of sleep and showed a dramatic increase in the duration and frequency of REM sleep episodes. The increase of REM sleep after deprivation is known as the REM rebound phenomenon. These findings suggested a need for REM sleep that, if thwarted, leads to a buildup of what has been called a REM pressure. In this sense, sleep scientists consider REM sleep to be a need or drive comparable to the conventional biological drives. If a person is deprived of a required substance such as water, the need increases, and when access to water is permitted, there is an increased consumption of it. However, in contrast to the deprivation of food, water, or oxygen, there are no obvious dire effects of depriving a person of REM sleep for a few nights.

Another direction of sleep research pioneered by Dement in the early days of REM sleep research is the animal work on REM sleep. In 1958, after completing his medical studies and while still a graduate student in physiology, Dement published a paper in "the *EEG Journal*" that described a stage of sleep in the cat when the EEG changed from the usual high voltage synchrony to an awake pattern of LVFA. Dement called this stage of sleep "activated sleep" (Dement, 1958). He also observed rapid eye movements and twitching movements of the legs, ears, and vibrissae during activated sleep and suggested that this phase of sleep in the cat is comparable to REM/dream sleep in humans that he and Kleitman had described in 1957. This line of research—that is, REM sleep in animals—was very soon recognized as an important development in sleep research and has since been studied in labs throughout the world.

In 1963 I attended a conference in New York City at which Dement was present. I had recently arrived at the University of Delaware and had started a sleep research program using cats. I mentioned to Dement that I had found his 1960 *Science* paper on REM sleep deprivation in humans very interesting and asked if he thought that REM sleep deprivation in cats was feasible. He thought so and agreed that the findings would be interesting. Encouraged by his response, one of the first projects I undertook at Delaware with a new graduate student, Tom Gordon, was such an experiment.

Cats were deprived of REM sleep for 3, 5, or 10 days, and a number of Dement's findings with humans were replicated in our cat experiment (Siegel and Gordon, 1965). On successive days of deprivation, dramatically more awakenings were necessary to prevent the cats from entering REM sleep. Ninety-four percent more awakenings were necessary on the last day of deprivation than on the first day. Our cats also showed the REM sleep rebound phenomenon. On the first recovery session when cats were permitted to have REM sleep, they immediately went into REM sleep and showed a 59% increase in the time spent in REM sleep compared with their usual time in REM sleep prior to deprivation. We concluded that cats, like humans, have a need or pressure for REM sleep. Since Dement had reported psychological effects of REM sleep deprivation, we used an emotional behavior rating scale for cats to evaluate any such changes (Norton and deBeer, 1956). We failed to see any changes in emotional behavior. After we submitted out manuscript for publication, a report by Danièle Jouvet-Mounier and her colleagues on REM sleep deprivation in cats appeared in a French journal (Jouvet-Mounier et al., 1964). Our findings were essentially the same as those reported in this and a more detailed paper by the same group published in 1966 (Vimont-Vicary et al., 1966).

12

The Neural Control
of REM Sleep

Shortly after the publication of Dement's paper on REM sleep in the cat, and for years following, Jouvet and a number of other workers demonstrated REM sleep in a variety of animal species (Jouvet et al., 1959; Jouvet and Valatx, 1962; Ruckebusch, 1962; Adey et al., 1963; Faure et al., 1963; Roldan et al., 1963; Hartmann et al., 1967; Shurley et al., 1969; Cicala et al., 1970; Schlehuber et al., 1974; Latash and Galina, 1975; Allison et al., 1977; J. M. Siegel et al., 1996, 1999). They showed REM sleep to be present in all mammalian species studied as well as in some other vertebrates, such as birds, but in very limited amounts. The discovery of REM sleep in animals led to two important developments. One was the recognition that this stage of sleep in humans is not just a curio associated with human dreaming and Freudian dynamics, but is a universal phenomenon in mammals. As such, REM sleep acquired the status of a legitimate topic of biological study—no longer relegated to the fringes of the psychoanalytical couch or turbaned dream analysts. The second development that followed the demonstration of REM sleep in animals was that the door was now open to invasive manipulations to explore the anatomical, physiological, and biochemical basis of REM sleep. With a very active research program including all these approaches, Jouvet soon became a dominant figure in sleep research.

Jouvet coined the term "paradoxical sleep" to refer to the REM sleep state, since the EEG displayed an awake rhythm but, paradoxically, the animal was behaviorally asleep. Other terms used for this state of sleep include "rhombencephalic sleep," referring to the hindbrain region found to be critical for it; "desynchronized" or "activated sleep" (Dement's term), referring to the EEG arousal pattern; and "dream sleep," referring to the reports of dreaming during this period. Some workers use "REM sleep" when referring to the human and "paradoxical sleep" in the animal case.

It is assumed by sleep workers, with good cause, that REM/dream sleep in humans and paradoxical sleep in animals are essentially the same physiological state. For humans, the most salient aspect of REM sleep is dreaming, but we do not know whether other animals actually dream. Many assume that animals do dream with whatever image-forming, cognitive, and symbolic capabilities they possess. Physiological indicators shared by humans and animals, in addition to rapid eye movements and an activated EEG, are an irregular heart rate and increased blood pressure, respiratory rate, and pupil diameter. These all indicate a heightened state of sympathetic activation—which is an integral part of emotional arousal. In addition, there is penile erection in males, as well as comparable signs of sexual arousal in females, which may point to a sexual component in the physiological arousal. Nonautonomic correlates of paradoxical sleep include transient spikelike potentials recorded from the pons, the lateral geniculate nucleus of the thalamus, and the occipital (visual) cortex. These are called "PGO spikes" or "PGO waves." They originate in the pons and spread synaptically to the geniculate and then to the occipital cortex. PGO spikes are associated in time with REMs and many agree that they are related to visual imagery in dreams.

Another nonautonomic component of REM sleep is loss of tonus in skeletal muscle, referred to as muscle atonia. This is rather dramatic in that it is a complete loss of tonus—more so than what one sees in an awake, relaxed state or even in quiet, deep slow-wave sleep. If electromyographic (EMG) electrodes are attached to a skeletal muscle group in humans, traditionally neck muscle under the chin, or to nuchal (dorsal neck) muscle in the cat, and recordings are taken during waking, one sees moderate to high amounts of muscle tonus depending upon the level of arousal. When the person or animal falls asleep, during slow-wave sleep, tonus decreases to a fairly low amplitude. When REM sleep is entered, the EMG trace falls to essentially zero, a flat line. If the amplification of the EMG channel is increased, nothing but amplifier noise is detected. This is comparable to what is seen if an animal is paralyzed with a neuromuscular blocking agent such as curare. It is a profound loss of tonus, essentially a paralysis of the postural, antigravity muscles. Fortunately, muscles of the diaphragm that control respiration are spared during sleep paralysis.

A final correlate of REM sleep is hippocampal theta. This is a regular and stylized EEG rhythm similar to the very regular 10 Hz alpha rhythm, but with a frequency of 4 to 8 Hz. Theta rhythm is generated by hippocampal neurons and is best recorded with an electrode directly in the hippocampus. However, in animals with a thin, lissencephalic cortex, such as the rat, theta can be recorded from the cortex, since the dorsal hippocampus theta generator is close to the surface of the cortex. In animals with thicker and convoluted cortices, like the cat, the hippocampus is not as close to the cortical surface and theta is not readily seen from surface recordings.

The significance of theta is a much-debated topic. It has been proposed to be related to emotional states, memory (especially spatial memory), and movement within one's spatial environment. Perhaps when workers on the hippocampus determine the significance of the occurrence of hippocampal theta, we may apply that to an understanding of theta occurrence during REM sleep. But at this time REM sleep theta, like PGO spikes, muscle atonia, and the autonomic changes, serves mainly as one of a number of physiological markers for the occurrence of REM sleep. Eventually, of course, we would like to understand the significance of the physiological changes that occur during REM sleep.

We talk about REM sleep as a unitary phenomenon, but given the various signs of REM sleep, we know that this is not accurate. In fact, REM sleep comprises quite distinct and very different aspects that should alert us to the possibility that there may not be a single generator that controls it. In the years following the discovery of REM sleep, concerted efforts were made to discover the brain region(s) responsible for it. By 1958, only five years after REM sleep was first reported, regions localized to the pontine reticular formation were discovered to be central to the control of the REM sleep state of arousal and its component parts. This is in contrast to the lack of finding a single region of the brain responsible for slow-wave sleep. The part of the pons involved in the control of REM sleep includes and is caudal to the area of the pontine–mesencephalic junction central to the waking form of arousal.

The introduction of variants of the *cerveau* and *encéphale isolé* preparations developed by Bremer in 1935 was instrumental in this discovery. In Moruzzi's lab in the late 1950s, Batini et al. (1958, 1959) experimented with transections of the brain stem at various levels between the classical *cerveau* and *encéphale* levels (see Figure 12.1A,D). Dramatically different effects were produced as a function of the level of the brain transection. This approach, pioneered in Moruzzi's lab, was

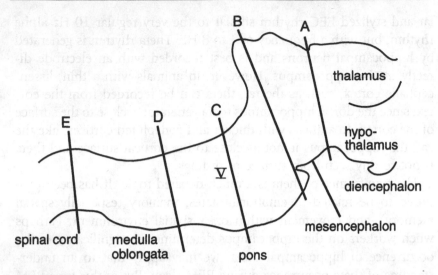

FIGURE 12.1. Sagittal view of the cat brain showing the locations of brain transections between the *cerveau* (A) and *encéphale isolé* (E) levels to parcel out the neural control of REM sleep. The roman numeral V between sections C and D indicates the location of the sensory and motor nuclei of the trigeminal (fifth) cranial nerve at the midpontine level.

explored further by Jouvet in Lyon, Zernicki at the Nencki Institute in Warsaw, and others. An extensive paper published by Jouvet in 1962 in the *Archives Italiennes de Biologie,* a journal edited by Moruzzi in Pisa, described the results of sectioning the brain stem at a number of levels between the diencephalic–mesencephalic junction and the medullary–spinal cord junction. Figure 12.1 schematically shows the levels at which the brain stem was transected to produce different effects.

Those who were considering the data from these preparations found it informative to look not only at the EEG signs of sleep and waking derived from cortical recordings that are in front of the transection, but to pay attention also to the signs of sleep and waking that can be detected from the head and body connected to and controlled by the brain *behind* the transection. That is, it was necessary to look not only at what goes on in front of the cut, but also at what is occurring behind the cut. This approach yielded valuable information.

A transection between the diencephalon and mesencephalon (essentially a *cerveau* cat, see Figure 12.1A) shows mainly slow-wave sleep recorded from the cerebrum, as first reported by Bremer and discussed earlier. Slow-wave sleep occurs here because the mesencephalic reticular arousal system is behind the cut and cannot influence the cerebrum. In contrast, the thalamic and other forebrain synchronizing structures that are in front of the cut are able to influence the cerebrum. When the mesencephalon was included with the forebrain (i.e., a preparation with a transection between the mesencephalon and pons: Figure 12.1B), the arousal function is added, since the reticular activating system can now exert its ascending arousal influence. However, forebrain recordings from this preparation did not show the signs of paradoxical (REM) sleep–type arousal. When attention was directed to what occurred *behind* the transection, it was discovered that parts of the head and body innervated by the nervous system behind the cut episodically displayed the signs of paradoxical sleep. These signs included muscle atonia, increased sympathetic arousal, and REMs (the cranial nerves controlling eye movements are behind this transection). In addition, the pontine component of PGO spikes was present in the pons, whereas the other components of PGO spikes, recorded from the thalamus and cortex, were not seen. Since the lower brain stem (the pons and medulla) was severed from the rest of the brain in front, the neural signals from the hindbrain required to produce the forebrain signs of paradoxical sleep were interrupted and, presumably, the cortex "did not dream" even though the body displayed its signs of paradoxical sleep. Since recordings from the head of an *encéphale isolé* preparation (Figure 12.1E) show signs of paradoxical sleep, we may conclude that the neural control of paradoxical sleep has nothing to do with the spinal cord and must reside in the brain stem somewhere between the rostral pons and caudal medulla (i.e., between transections B and E in Figure 12.1).

The transections that proved most illuminating were through the brain stem at the midpontine level just anterior to the sensory and motor roots of the trigeminal nerve, called a midpontine pretrigeminal section (Figure 12.1C), and a transection between the pons and medulla (Figure 12.1D). Signs of paradoxical sleep still did not appear rostral to the midpontine pretrigeminal section; based on these data, the rostral half of the pons was ruled out. A section at the pontine–medullary junction shows complex effects that have been studied in detail by Jerome M. Siegel at UCLA. In a paper entitled Ponto-Medullary Interactions in the Generation of REM Sleep, Siegel (1985) described brain stem transection studies in conjunction with single-unit recordings from brain stem neurons involved in the control of paradoxical sleep. These findings, in addition to the earlier work, led to the conclusion that the major brain control of paradoxical sleep resides in the mid- to caudal pons. Siegel also showed that the medulla is involved in the control of muscle atonia and in determining the duration of paradoxical sleep episodes. The role of the medulla in the timing of paradoxical sleep has been supported by findings of Jouvet (Jouvet et al., 1995).

To locate a region in the pontine brain stem specific for the control of paradoxical sleep, discrete lesions, focal electrical and chemical stimulation, and recording experiments have been made in the pons. As might be expected from knowledge that paradoxical sleep is composed of a number of components (REMs, EEG desynchronization, muscle atonia, etc.), a number of areas within the pons appear to be responsible for certain of the individual components. We will limit the discussion here to the most salient descending component of paradoxical sleep, muscle atonia, and to the main ascending component, cortical EEG activation.

The evidence that the pontine brain stem is responsible for the atonia in paradoxical sleep, independent of the other signs of paradoxical asleep, was surveyed by Pompeiano in 1976. In 1979 Jouvet's lab reported that electrolytic lesions to a small population of cells in the dorsolateral pons, the nucleus subcoeruleus, adjacent to the locus coeruleus, led to the selective loss of muscle atonia during paradoxical sleep (i.e., muscle tonus remained during paradoxical sleep) (Sastre and Jouvet, 1979). A detailed study of the effects of dorsolateral pontine lesions on muscle tonus during paradoxical sleep in cats was reported in 1982 by Adrian Morrison and his colleagues at the University of Pennsylvania (Hendricks et al., 1982). Interesting aspects of these reports are the descriptions of the behavior of animals with such lesions, which also provide an understanding of the utility of muscle paralysis during paradoxical sleep. With such lesions, after a period of time in slow-wave sleep, when the animal entered paradoxical sleep, as indicated by the standard physiological changes, the cat appeared to be acting out a dream. Morrison filmed cats with such pontine lesions, and the behavior is most exotic. As expected from the signs of strong sympathetic arousal, behaviors released were of a full-blown emotional nature. Cats that had been sleeping quietly and peacefully would suddenly get up and behave as if they were aggressively attacking something or escaping some awful thing, even though they were oblivious to actual objects presented to them. Of course, nothing of a threatening or provocative nature was in the cage with the cat. This behavior is quite strange to see.

The issue of how the dorsolateral pons affects muscle tonus during paradoxical sleep has yet to be considered. Since atonia occurs in the midpontine pretrigeminal preparation, it must be a *descending* influence from the pons; that is, it cannot be mediated by a pontine inhibitory influence acting upon upper motor neurons of the cerebral cortex. Michael Chase and his colleagues at UCLA conducted a series of remarkable studies, a technical tour de force, in which they recorded intracellular potentials from brain stem and spinal motoneurons in unanesthetized cats during normal sleep and waking. Their work provided evidence that muscle atonia during paradoxical sleep is due to a projection from the nucleus subcoeruleus to a region in the medulla, the nucleus gigantocellularis, which in turn projects to the spinal cord. During paradoxical sleep, these medullary neurons, via the inhibitory transmitter glycine, exert a descending tonic inhibitory effect, membrane hyperpolarization, upon the final motor neurons of the spinal cord that control skeletal muscle (Chase and Morales, 2000).

EEG desynchronization is a most telling feature of paradoxical sleep. It signifies a cerebral cortex that is cognitively active; it is the sine qua non of the dreaming brain. As we learned earlier, with the midpontine pretrigeminal preparation we can "dissect" this stage of sleep in the sense that the parts of the animal still connected to the pons show the signs of paradoxical sleep and the parts disconnected do not. The body and regions of the head controlled by cranial and spinal nerves behind the cut show paradoxical sleep (muscle atonia, REMs, and sympathetic activation); the cerebral cortex in front of the transection does not. The cerebrum is not dreaming, but the rest of the body, in a physical sense, is. Which region of the pons is responsible for the desynchronization of the cortex seen in paradoxical sleep?

Evidence on this issue has been generated by a number of laboratories; I will focus on the major findings of a few. On this topic, particularly, the actions of various neurotransmitters play a dominant role. Recordings from cells of the pedunculopontine (PPT) and laterodorsal tegmental (LDT) nuclei, which are predominantly cholinergic, showed that the majority of PPT cells increased their firing rate about a minute prior to the very first sign of paradoxical sleep—suggesting that this nucleus has a primary role in the generation of paradoxical sleep (Datta et al., 1989; Steriade and McCarley, 1990; Steriade et al., 1990; Datta and Hobson, 1994). Datta (1995) has reviewed the evidence for the involvement of nuclei PPT and LDT in REM sleep.

These two nuclei, collectively called the "peribrachial nuclei," were described earlier in Chapter 5, on arousal. Figure 5.1 in that chapter shows the location of these nuclei in the rostral dorsal pontine tegmentum. To qualify as a structure that can desynchronize the cortex, we would expect it to project to the thalamic nuclei that control cortical rhythms. In Chapter 8, on the neural circuits that produce sleep, there is a description of the role of the peribrachial nuclei in controlling the shift from slow-wave sleep to waking. In that context, we discussed the projection of the cholinergic neurons to a number of nuclei, including the reticular and intralaminar nuclei of the thalamus, to disrupt the low frequency synchronogenic drive to the cortex and produce an activated cortex.

We see now that the peribrachial nuclei are also involved in the cortical activation seen during REM sleep. Are these two forms of cerebral activation the same? Looking at the cortical EEG, it appears that the two states of arousal are similar. EEGs recorded during both states show similar patterns of frequencies and amplitudes. The often-stated view is that cerebral arousal during waking and during REM sleep are essentially the same. However, an obvious difference is that cognitive activity during waking is based on the processing of external information, whereas during REM sleep internal information is involved. Another difference, based on recent evidence from brain imaging studies, is that the subcortical structures and pathways involved in ascending arousal are different during waking and during REM sleep.

Positron emission tomography (PET) has produced images of the brain during REM sleep arousal demonstrating that limbic and limbic-related regions of the forebrain are prominent users of metabolic resources (Maquet et al., 1996; Braun et al., 1997; Nofzinger et al., 1997). It has been known for a long time that the limbic system plays a major role in controlling emotional behavior. Given the nature of the autonomic changes that occur during REM sleep, the emotional display of cats with pontine lesions that abolishes REM sleep paralysis, and the prominence of emotional content in dream reports, it is not surprising that the PET studies showed limbic system activation during REM sleep. Since the EEG is relatively insensitive for showing differences among brain structures during arousal in the waking and REM sleep states, it is of value to have the PET data to indicate where and what the differences are.

Based on PET data showing the brain regions that are active during REM sleep in comparison to the regions active during waking, Braun et al. (1997) proposed that during REM sleep, ascending activation that originates in the brain stem reticular system is mediated by a ventral pathway through the basal forebrain. (The ventral and dorsal pathways for ascending activation of the cortex were described in Chapter 5 and depicted in Figure 5.2.) The ventral route results in a basal forebrain, cholinergic activation of the cerebral cortex for a REM sleep/dream-type arousal. This route also engages hypothalamic and limbic structures, which would account for the emotional aspects of dreams. This is in contrast to the dorsal route from the reticular system that activates the cortex during waking. The dorsal route is mediated by the intralaminar and midline nuclei of the thalamus, which produce a glutamatergic activation of the cortex. The dorsal pathway, active during waking arousal, may be responsible for waking cognitions that are ordinarily well organized and logical—in effect, reality oriented. These qualities of consciousness are suspended during dreaming. In emotional states during waking, it would appear that the ventral as well as the dorsal pathway is involved, with the dorsal pathway providing somewhat of a logical check to the emotional response.

We note that the cholinergic neurons of the pons that control the ascending activation of the cerebrum during paradoxical sleep are not in the caudal part of the pons but in the rostral pons. In fact, based on data from electrophysiological studies, Steriade refers to the region as the "mesopontine junction" (Steriade, 1996). A possible source for the discrepancy between this and the conclusion from transection studies, which pointed to the mid- and caudal pons, is that whatever instrument is used to make the transection (e.g., a fine spatula passed through the brain stem to sever it at the desired level) unavoidably causes edema and damage to adjacent neural tissue that is rostral and caudal to the cut. Thus, the spatial resolution of the location of areas controlling particular functions is limited. In contrast, electrical stimulation with fine-tipped electrodes or chemical stimulation via pipettes for infusion or iontophoresis are capable of affecting very localized areas of brain tissue. An added advantage of chemical stimulation is that it can selectively target neurons with particular receptor sites and avoid activating fibers of passage, which is a problem with electrical stimulation.

Another technique for precise localization is that of recording from single cells with microelectrodes. After the functional properties of neurons in an area have been recorded, a dye is deposited or a small marking lesion is made; subsequent histological analysis will show the precise location of the cells from which recordings were obtained. These techniques, used by workers such as Steriade, point to the rostral pons (the peribrachial nuclei) for the control of ascending cortical desynchronization and to the middle and caudal pons as control areas for other aspects of paradoxical sleep. All in all, we are not talking about a great discrepancy between the transection and electrophysiological evidence, only about a millimeter or two in the cat brain.

We have described a number of brain stem structures, mainly at the pontine level, that generate the individual signs of REM sleep. A conclusion reached by workers in this field is that the different brain stem regions that control the individual components of REM sleep are interconnected. A fair amount of effort has been and still is directed to discovering the pathways and neurotransmitters that interconnect these brain stem structures.

Another issue related to the control of paradoxical sleep (besides the brain structures involved) is the timing of the episodes. Clearly, its occurrence displays a species-specific cyclicity. In the human, REM sleep occurs about every 90 minutes; in the cat the cycle time is a little under 30 minutes. There are a number of neural mechanisms that can produce rhythmic cycling. One we are most familiar with is the circadian rhythm governed mainly by cells of the suprachiasmatic nucleus of the hypothalamus, which possess an intrinsic rhythm of membrane excitability. There are also pacemaker cells in the reticular nucleus of the thalamus. However, there is no evidence for such cells in the pontine brain stem region, which is important for REM sleep.

Cycling may also be achieved by a reciprocal interaction between cells located in separate nuclei. In the case of REM sleep, one population of cells would control the onset and the other would control the offset of the REM sleep episode. An elegant version of this approach was proposed by Allan Hobson and Robert McCarley of Harvard University in two papers published in 1975 in the journal *Science* (Hobson et al., 1975; McCarley and Hobson, 1975). The first paper (Hobson et al., 1975) describes, in neurophysiological terms, the theory and the neural data supporting it. The companion paper (McCarley and Hobson, 1975) is quite novel and creative. It presents a mathematical model, taken from the field of animal ecology, that describes the reciprocal interactions between predator and prey populations that result in a stable oscillation of the sizes of their respective populations. McCarley and Hobson apply the equations of this model to the firing rates of two populations of nerve cells that, at the time, they considered to be the control nuclei for the onset and offset of REM sleep. They demonstrate how the equations for reciprocal interactions yielded theoretical firing rates that closely approximated the cyclical alterations of firing rates seen in two populations of neurons recorded from the cat pontine brain stem at the onset and offset of REM sleep.

Their initial presentation of the theory incorporated experimental evidence available at the time about the brain stem nuclei that control REM sleep. As additional information was generated in subsequent years, the theory was refined and revised, most notably in the book coauthored by Steriade and McCarley (1990) in which the neurons that control the onset of REM sleep are identified as being the cells of the LDT and PPT, the peribrachial nuclei. More recently, Hobson et al. (1998) have altered and extended the theory.

An aid to the understanding of this theory is shown in Figure 12.2, a diagram based on the early description of their model (McCarley and Hobson, 1975) with additional information provided in later descriptions of the model (Steriade and McCarley, 1990; Hobson et al., 1998). Since the1970s, specific brain nuclei and neurotransmitters have been identified with the control of REM sleep onset (REM-on) and the termination of REM sleep (REM-off). I will present a current view on this in the context of the Hobson and McCarley model, noting that there is not total agreement on this by all workers in the field and that the story is still evolving.

As seen in Figure 12.2, REM sleep occurs when the inhibitory influence from the REM-off region is removed. In this sense, the REM-off neurons, when not active, play a permissive role for REM sleep. Experimental evidence for this is that the noradrenergic cells of the locus coeruleus and serotonergic cells of the dorsal raphé nucleus show a dramatic decrease and close to zero firing rate just prior to and at the onset of REM sleep. These aminergic neurons (norepinephrine and serotonin are biogenic amines) have widespread projections, with some projecting a long distance to the cerebral cortex and others a short distance to neighboring neurons. Relevant to this discussion are the projections to the nearby peribrachial nuclei, PPT and LDT of the mesopontine area. These aminergic neurotransmitters have an inhibitory effect on the cholinergic cells of the peribrachial nuclei. When the locus coeruleus and raphé neurons decrease their firing, their inhibitory influences are removed, *permitting* the peribrachial cells to increase their firing rate (see influence A_1 in Figure 12.2). This disinhibition may thus be seen as having a *permissive* effect. As the firing rate of aminergic neurons slows down and permits a reciprocal increase of firing of the peribrachial neurons, a critical level of activity is reached that triggers the various aspects of REM sleep. The signs of REM sleep are mediated by the increased release of acetylcholine from the terminals of PPT and LDT neurons that, like the locus coeruleus and raphé systems, have widespread projections (not represented in Figure 12.2). Basic to the cerebral arousal seen in REM sleep is the peribrachial cholinergic projection to the nucleus basalis of the basal forebrain, which in turn has a diffuse cholinergic projection to the entire cortical mantle.

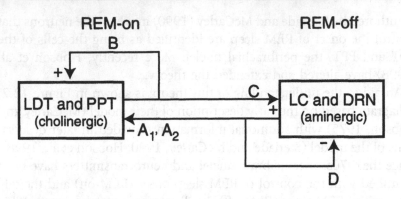

FIGURE 12.2. Diagram of the McCarley and Hobson model of the neural control of REM sleep. The onset of REM sleep (REM-on) is initiated by decreased activity of aminergic neurons of the locus coeruleus (LC) and the dorsal raphé nucleus (DRN). These aminergic neurotransmitters (norepinephrine and serotonin) have an inhibitory effect on the cholinergic cells of the peribrachial nuclei, the lateral dorsal tegmental (LDT), and the pedunculopontine tegmental (PPT) nuclei (influence A_1). When the aminergic neurons decrease their firing, cells of the LDT and PPT are released from their inhibition, and the peribrachial cells increase their activity. The release of acetylcholine from these cells triggers the onset of REM sleep by activating brain regions that control the various components of REM sleep (not illustrated). The REM-on process is facilitated by short collateral fibers of the cholinergic projection neurons from LDT and PPT that produce a positive feedback onto autoreceptors to increase the activity of these cells (influence B). Cholinergic projection neurons have an excitatory effect on the aminergic cells of LC and DRN (influence C). As REM sleep continues, this excitation will be increasingly effective in driving the activity of LC and DRN cells, which in turn will inhibit the REM-on cells (influence A_2) and so terminate the REM sleep episode. Short collateral fibers from the aminergic projection neurons project back to their cells of origin to produce feedback inhibition (influence D) and, after a given amount of negative feedback, remove the feedforward inhibition (A_2) and permit another REM sleep period. (McCarley and Hobson, 1975, p. 59. Adapted with permission from McCarley RW, Hobson JA (1975) Neuronal excitability modulation over the sleep cycle: a structural and mathematical model. *Science* 189:58–60. © 1975 American Association for the Advancement of Science.)

The drive into REM sleep by the cholinergic peribrachial neurons is furthered by at least two processes. One is the decreased inhibition of the peribrachial cells due to the reduced firing of the aminergic inhibitory cells. This is the permissive role just described. In addition to reduced inhibition, there is an excitatory influence on the peribrachial cells. Peribrachial cells possess short collateral axons that feed back onto their own cholinergic autoreceptors to increase the excitation and activity of these cells (B in Figure 12.2). This is a process of feedback excitation. There is evidence that these cholinergic neurons also have local excitatory reticuloreticular connections to other regions of the brain stem (not shown in Figure 12.2) to further the drive into the REM sleep state and to orchestrate the other signs of REM sleep (e.g., PGO spikes, REMs, muscle atonia).

A nearby region of the brain stem to which the cholinergic neurons of the peribrachial nuclei project is just ventral and medial in the rostral pontine reticular formation. This region of the reticular formation has a number of names, one of which is nucleus pontis oralis, or NPO ("oralis" means rostral; there is also a nucleus pontis caudalis, NPC). There is evidence that cells of the NPO send an excitatory amino acid projection to activate the cholinergic basal forebrain. The basal forebrain receives afferent projections from the peribrachial nuclei as well as the NPO, both of which activate the cholinergic basal forebrain projection to the cerebral cortex to produce EEG desynchronization.

Recent work by Chase and colleagues (Xi et al., 1999) shows an important role of γ-aminobutyric acid (GABA) in mediating the control of NPO neurons to produce the desynchronization of waking and of REM sleep. When GABA receptors of NPO cells were pharmacologically activated, EEG desynchronization and behavioral wakefulness were produced. When the GABA receptors were blocked, REM sleep resulted. It appears that turning on the GABA inhibitory influence to the NPO results in wakefulness. In contrast, turning off, or inhibiting the GABA influence—thus releasing NPO cells to be active—results in REM sleep. The REM-on system appears to be controlled by the joint action of the peribrachial nuclei and nearby NPO. The exact nature of the interaction between these neighboring pontine nuclei and the anatomical source of the GABA terminals in the NPO is not known.

As REM sleep continues, the above-mentioned processes in the cholinergic nuclei of the pons will result in an excitatory influence to the REM-off neurons of the locus coeruleus and raphé dorsalis (C in Figure 12.2). The activity of the REM-off cells will increase, and these neurons become increasingly effective in inhibiting the activity of the REM-on neurons (A_2 in Figure 12.2), terminating the REM sleep episode. But built into the increased activity of the aminergic neurons is their negative feedback, as shown in D of Figure 12.2. This will contribute, after a given amount of negative feedback inhibition, to the removal of the feed-forward inhibition of the REM-on cells and so permit the resumption of another REM sleep episode.

There remains the issue of what controls the timing of the REM sleep cycle, that is, the interval between the onset of successive REM sleep episodes. Since there is no evidence of circadian-like intrinsic oscillator cells in the brain stem region to control REM sleep cycles, Steriade and McCarley state that REM sleep oscillations must be controlled by interactions between neuronal populations (Steriade and McCarley, 1990, p. 367). These authors do, however, invoke the diencephalic circadian oscillator (the suprachiasmatic nucleus) for the initial turning on and final turning off of the REM sleep episodes. They state that the circadian oscillator interacts with the REM sleep oscillator and other influences to accomplish that.

In his book *The Dreaming Brain,* Hobson suggests that the REM sleep cycle time is governed not by neural events on the time scale of action potentials and postsynaptic potentials that occur on the order of milliseconds, but by cellular events that occur over many minutes and longer (Hobson, 1988, pp. 191–192). He proposes that when the neurons of the REM-on and REM-off nuclei receive signals to be active, the cells transform the brief message into a slow biochemical process. The neural input signal initiates rapid biochemical and electrical events in the cell body for the generation of the postsynaptic action potential. That process is on the order of milliseconds. The cell body has the biochemical machinery and the enzymes for neurotransmitter synthesis. Transmitter molecules must then be transported down the axon to axon terminals for storage and subsequent release by the fast-moving action potentials. Axoplasmic transport occurs at slow and rapid rates (in the range of less than 1 mm/day to 400 mm/day). Hobson invokes a rate of transport and combines that with the distance between the relevant control nuclei to arrive at the cycle time for a given species. This ingenious hypothesis would be interesting to test.

Other modifications of Hobson and McCarley's 1975 theory are based on evidence that the REM-on and REM-off nuclei are not exclusively cholinergic and aminergic, respectively. In addition, REM-on and REM-off feedback effects may be mediated by excitatory and inhibitory interneurons that utilize glutamate and GABA, respectively, and other transmitters and neuromodulators also impinge upon the two components of the oscillator. But the basic position of the model appears to be firmly established: a mainly cholinergic control of REM sleep onset via REM-on neurons of the peribrachial nuclei and a mainly aminergic control of REM sleep offset by REM-off neurons of the locus coeruleus and nucleus raphé dorsalis.

Part III
The Functions
and Disorders of
Sleep and Waking

13

Theories of Sleep and Waking

The utility and function of waking are obvious. This is the state during which we eat, drink, procreate, create, and recreate. The survival value of these behaviors for the individual and for the species is also obvious (at least for the first three behaviors listed). In contrast, an understanding of the utility and function of sleep is not so easy.

The function of sleep, at one level, seems simple. After being awake and active during the day, we become tired and cannot function effectively. Many of us know that a night of lying in bed and not sleeping does not refresh us as if we had slept, but if we sleep for a number of hours, we awaken refreshed and can function effectively again. It is obvious, at this level, that sleep has restored us. As suggested by Shakespeare's line quoted in the preface of the book, "Thy best of rest is sleep," more than just rest is required. What happens at the cellular level in the brain or elsewhere to accomplish this marvelous effect is not known. There are theories of course, and considerable research is being done to provide answers to the "how" and "why" of sleep.

Evolutionary and Ontogenetic Observations on Sleep and Waking

A tantalizing observation about the wake/sleep cycle is that all animals, from single-celled organisms to human beings, show periods during which they are active and periods of behavioral quiescence. The quiescent (rest) periods in mammals and birds can, by a number of criteria, be described as sleep. Some workers are willing to include reptiles, amphibians, and fish as animals that sleep. The criteria for a period of rest to be considered as sleep include the following (Campbell and Tobler, 1984; Hendricks et al., 2000; Shaw et al., 2000):

1. Circadian controlled cycles: the episodes of rest and waking normally occur during particular periods of the 24-hour circadian cycle.
2. A species-specific posture: the animal assumes a characteristic species-specific posture when it enters the resting state.
3. An elevated threshold for arousal: a stronger sensory stimulus is required to elicit an orientation response when the animal is in the quiescent state as opposed to being clearly awake.
4. Rebound after deprivation: when an animal is permitted to enter the resting state after being prevented from doing so for a period of time, an increased amount of time will be spent resting. This implies that a homeostatic mechanism regulates a normal amount of time to be spent in the resting state. If less than that amount of time is permitted (sleep deprivation), the homeostatic control mechanism is activated and the rebound occurs when rest is permitted during a recovery period.

Some recent studies using insects have yielded fascinating data that show remarkable similarities between sleep, as we know it, and periods of rest in the fruit fly and the honey bee. To paraphrase an old Cole Porter song, and a newspaper report in the Los Angeles Times of January 4, 2001: rats do it, cats do it, birds do it, dolphins and people do it, and now we know fruit flies and bees do it—sleep, that is. The data for both insects are similar; I will describe the fruit fly work.

This common insect (*Drosophila melanogaster*), long a favorite organism for genetic studies, has in recent years been used in behavioral studies. Fruit flies have been shown to be capable of learning a variety of tasks, and workers have been exploring the genetic mechanisms of learning. In very recent times, two laboratories have published papers showing sleeplike behavior in *Drosophila*. Their titles tell the story: Correlates of Sleep and Waking in *Drosophila melanogaster* (Shaw et al. 2000) and Rest in *Drosophila* Is a Sleep-like State (Hendricks et al. 2000).

Using a variety of simple and ingenious techniques, both groups of workers showed that periods of rest occurred predominantly during the 12-hour dark phase of a 24-hour period (a circadian rhythm of rest and waking). After a period of rest deprivation, there was a robust rebound effect, with increased time spent in the rest state. The amount of rest during the rebound was about 50% percent of the rest that was lost during the deprivation period (Shaw et al., 2000). That is comparable to the amount of sleep rebound seen in mammalian sleep deprivation. Both studies showed elevated arousal thresholds during periods of behavioral quiescence. Hendricks et al. also reported that during periods of rest and increased arousal thresholds, the flies were immobile and usually in a prone position (a species-specific posture).

An interesting and provocative finding of both studies was that when caffeine, a known stimulant in mammals, was administered to *Drosophila*, the amount of time spent resting decreased. This was shown to be a dose-dependent effect: the higher doses of caffeine resulted in a greater reduction of rest. The sleep-reducing effect of caffeine is known to be due to its antagonistic effect on adenosine receptors. Since it is the adenosine A_1 receptor that is relevant for the sleep-reducing effect of caffeine, Hendricks et al. administered a selective A_1 receptor agonist and found that the duration of rest was significantly increased. In Chapter 8, on the neural mechanisms that produce sleep, the work of McCarley on the role of adenosine was described. Adenosine, a product of cellular metabolism, produces sleep via its action on adenosine A_1 receptors located in specific regions of the mammalian brain. The same neurochemical has essentially the same effect in an insect with a primitive nervous system and on the complex nervous system of a mammal. This is a prime example of what biologists refer to as the process of conservation in evolution. Some of the biochemical mechanisms that evolved in early primitive organisms are maintained (conserved) throughout evolution. It appears that a neurochemical that regulates rest/sleep in the fruit fly has been conserved through evolutionary history to serve essentially the same function in advanced complex organisms.

These findings show that a sleeplike state exists in animals as primitive as insects. A question of interest now is, At what point in evolutionary history did REM and slow-wave sleep emerge as two distinct states of sleep? This issue has been studied by using living animals that represent evolutionarily ancient species from which modern mammals emerged, that is, studies conducted on the most primitive living mammals.

The class of mammals consists of three orders: placental mammals (e.g., rats, cats, humans); marsupial mammals, which carry their young in pouches (e.g., kangaroos); and monotremes, the most primitive of mammals. Monotremes, like reptiles and birds, lay eggs, but like mammals, they secrete milk. The ancient ancestors of extant monotremes may represent the evolutionary link between reptiles and birds, and mammals. Placental mammals have been most successful at reproducing and surviving. There are about 4000 extant placental mammalian species and 300 marsupial species; except for three species, the order of monotremes is extinct. The extant species of monotremes are indigenous to Australia, Tasmania, and New Zealand.

In 1972 Allison and colleagues studied sleep in one of the three species of monotremes, the short-beaked echidna, commonly called the spiny anteater. They readily observed waking and non-REM sleep. During quiescent periods after a time in non-REM sleep, they saw signs that they tentatively identified as indicators of REM sleep. However, after extensive electrophysiological and behavioral testing, they concluded that what they had suspected was REM sleep was a state of quiet waking. Allison and coworkers reported that this primitive mammalian species experienced only two states of consciousness, waking and non-REM sleep (Allison et al., 1972).

Sleep in the monotreme order of mammals has been revisited more recently. In studies published in 1996 and 1999, J. M. Siegel, at UCLA, in collaboration with coworkers from UCLA and the University of Queensland in Brisbane, Australia, described sleep in the short-beaked echidna and in the duck-billed platypus (J. M. Siegel et al., 1996, 1999). The research was done in Australia, the natural habitat of these primitive mammals. Many of the findings on the echidna (the 1996 paper) concurred with those of Allison et al. (1972). During periods of behavioral quiescence and low voltage EEG, there was an absence of rapid eye movements. Similarly, as with the findings of Allison and his colleagues, brain temperature changes and arousal thresholds during periods the researchers thought might be REM sleep indicated that REM sleep, *as we know it* in placental mammals, is not present in this species of monotremes.

The significant and interesting contribution in the 1996 paper by Siegel et al. is based on the use of single-cell recordings. J. M. Siegel and others had found in earlier work that nerve cells in the brain stem reticular formation of placental mammals display different patterns of firing rates and variability of firing rates during waking, slow-wave sleep, and REM sleep (Huttenlocher, 1961; Hobson et al., 1974; J. M. Siegel et al., 1977, 1979; J. M. Siegel and Tomaszewski, 1983; J. M. Siegel, 2000b). During active waking, firing rate and variability were high; during quiet waking, rate and variability were reduced; during non-REM sleep, rate and variability were further reduced to a minimum. Distinctively, during REM sleep, rate and variability increased dramatically. During REM sleep, cells in the brain stem displayed rates and variability of firing that resembled the firing pattern seen during active waking. J. M. Siegel et al. (1996) found that brain stem single-cell discharge patterns recorded during active and quiet waking in echidna resembled those seen in placental mammals during similar behavioral states. During the echidna sleep state, which resembles slow-wave sleep in other mammals, the discharge rate of single cells decreased, but the variability of firing rate *increased*, as seen in REM sleep in other mammals. The increased variability seen in echidna was intermediate between the variability seen during slow-wave sleep (low) and during REM sleep (high) in cat and dog brain stem neurons. In other words, the firing pattern of brain stem cells in echidna during sleep tends toward, but is not quite like, the firing pattern of brain stem cells of placental mammals during REM sleep. Siegel and his coworkers concluded that the single state of sleep in this primitive mammal displays some of the characteristics of both slow-wave and REM sleep.

J. M. Siegel et al. (1996) refer to an observation of Corner and Bour (1984) that the pattern of brain stem neuronal activity seen in the adult placental mammal is gradually arrived at during a period of postnatal development. Siegel et al. (1996) then point out that the neuronal firing pattern of the adult monotreme mammal resembles that of the neonate placental mammal. They suggest that the differentiation of sleep into REM and slow-wave sleep during early development in the placental neonate recapitulates the phylogenetic differentiation of REM and slow-wave sleep from a primordial sleep state as seen in echidna.

In 1999 Siegel and the same colleagues with whom he had worked on the echidna, published a report on another of the three species of monotremes, the duck-billed platypus, a semiaquatic animal (J. M. Siegel et al., 1999). This platypus is descended from an ancient platypus-like ancestor from whom the short- and long-beaked echidnas are thought to be derived. The living platypus is considered to be the most primitive extant mammal. Since there were brain stem signs of REM sleep in the short-beaked echidna, it was of interest to see what, if any, signs of REM sleep might exist in this most primitive monotreme.

The lowest amplitude EEG recordings occurred when the platypus was awake. During all periods of quiescence and behavioral sleep, the EEG showed higher amplitude voltages. When the animal was awake and quiet, muscle tonus was reduced or, in some cases, totally absent. So two of the classic indicators of REM sleep in placental mammals (low voltage EEG and muscle atonia) were not useful in the platypus. When it entered a state of quiet sleep, the EEG amplitude increased somewhat. In some cases of sleep onset, after only 30 to 90 seconds, the animal exhibited episodic rapid twitching movements of the head, neck, and bill, as well as rapid eye movements. These clear indicators of REM sleep were accompanied, however, by EEG recordings of moderate to high amplitudes. Episodes of phasic twitching and rapid eye movements occupied only 4 to 6% of the time during periods of high voltage sleep. However, during periods of moderate EEG amplitude, this REM phase occupied almost 50% of the sleep time. It appears that a REM-like state occurs during sleep when the EEG amplitude is moderate, and there is relatively little REM sleep when the EEG voltage is of high amplitude. Siegel and his colleagues concluded that this most primitive of mammals has REM sleep that is limited to the brain stem, and that the forebrain activation (low voltage EEG) seen in advanced mammals does not occur in monotremes. These data suggest that forebrain activation in placental animals during REM sleep is the neural correlate of cognitive/mental phenomena, and the brain stem activation seen in all mammals is related to the physiological/peripheral components of REM sleep (atonia, muscle twitches, and rapid eye movements).

An interesting additional observation is that the platypus spends about 50% of total sleep time in REM sleep. This is, by far, more time in REM sleep than is seen in other adult animals. This finding is comparable to the one noted about the short-beaked echidna: namely, the adult echidna displays a firing pattern of brain stem neurons similar to that seen in neonate placental mammals. So in these two species of monotremes, we see evidence for ontogeny recapitulating phylogony. At least one intriguing pair of questions remains to be answered: What adaptive function do these aspects of REM sleep serve in the evolution of these most primitive mammals? And what similar or related function do they serve in placental mammals?

We consider now another characteristic of sleep that appears to be highly significant, if only we can interpret it correctly. This is the ontogenetic development of sleep in mammalian species. In animals born in a fairly primitive state in which the nervous system undergoes considerable further development after birth, the amount of REM sleep is dramatically high at birth. Jouvet-Mounier et al., in 1970, reported that neonate rats and cats, both with primitive nervous systems at birth, spend between 90 and 95% of total sleep time in paradoxical sleep. (The custom in Jouvet's lab is to use the term "paradoxical sleep" rather than "REM sleep.") Within a month after birth, the amount of paradoxical sleep decreases to its adult level of approximately 20% of total sleep time. In animals that are born fairly well developed, such as ungulates (e.g., sheep and cattle) that can walk at birth, the amount of paradoxical sleep at birth is close to what it is in the adult. Much neural development along with large amounts of paradoxical sleep occur *prior* to parturition in these animals. Humans are intermediate in this respect; about 50% of total sleep time is spent in REM sleep at birth. Note that neonate humans sleep between 16 and 20 hours per day. That translates into 8 to 10 hours of REM sleep per day! Contrast that with adults, who spend about 20% of total sleep time in REM sleep (1.5 hours of REM sleep per day). If the EEG is monitored in utero at seven months, the finding is that REM sleep occupies about 80% of sleep time. That works out to be 16 hours per day in REM sleep! What do these data suggest?

In 1966 Roffwarg, Muzio, and Dement published a seminal paper in which they proposed a theory of REM sleep based on ontogenetic findings (Roffwarg et al., 1966). They refer to evidence that development and myelinization of the fetal and neonatal nervous system is facilitated by neural activity. They note that neurons in the fetal brain show development and myelinization in areas related to tactile, proprioceptive, auditory, and gustatory pathways—modalities in which there is considerable sensory stimulation in the intrauterine environment. However, the visual system that is without exogenous stimulation also shows substantial myelinization prior to birth. Roffwarg et al. suggest that the neural activity in the visual system related to the occurrence of PGO spikes during the REM sleep in utero is responsible for this. Furthermore, they propose that the nervous system in utero is altered functionally as well as structurally by REM sleep. They suggest that neural circuits of the fetal and neonate brain are activated during REM sleep in patterns required to serve organized behavior later in life. These patterns of neural activity bring the "higher brain centers to an operational capacity requisite to handling the rush of external stimulation in waking experience" (Roffwarg et al., 1966, p. 616). Jouvet has expressed a similar position: "Paradoxical sleep may be genetically programmed *in utero* . . . to stimulate and organize the delicate synaptic circuitry which is needed for some instinctive behavior which has to be ready immediately after birth" (Jouvet, 1972, p. 270).

This attractive hypothesis about the function of REM sleep was proposed over 30 years ago, and we are only minimally closer now to verifying that position than we were then. Studies that support the theory show that increased neural activity facilitates brain development (Zheng and Purves, 1995)—and in the fetal brain increased rates of neural activity occur during the REM state (Mirmiran and Van Someran, 1993;

Mirmiran, 1995). Specifically related to development of the visual system, it has been shown that REM sleep deprivation or the abolition of PGO spikes in neonatal cats impairs the normal development of neurons in the visual nucleus of the thalmus, the lateral geniculate body (Davenne and Adrien, 1984; Pompeiano et al., 1995; Oksenberg et al., 1996). Much of the experimental evidence on the functions of sleep is based on sleep deprivation studies in adult animals and humans. These are described next.

Sleep Deprivation Studies

Early examples of the sleep deprivation approach were described in Chapter 9. To assure the buildup and detection of a chemical that might be called a sleep factor, animals were deprived of sleep for a number of days (sufficient to harvest a chemical factor or factors from blood, cerebrospinal fluid, or brain tissue itself). Sleep deprivation did not cause dramatic or obvious behavioral changes, except the apparent urge to sleep. Subsequent experiments used careful behavioral testing to detect sleep deprivation effects. Sleep deprivation studies fall into one of three categories: partial deprivation (restricted amounts of sleep per night), short-term deprivation (one to four nights without sleep), and long-term deprivation (five or more nights of sleep deprivation).

Partial sleep deprivation, involving reduced amount of sleep per night, is of some popular interest based on the notion that we have so much to do and are hard-pressed to do it all during our waking hours. If we can reduce sleep time by just a few hours per night, who knows how much more we could accomplish! Students can do this for a few days, for example, at the end of term when exams and term papers must be dealt with, but it does not seem to be possible for an extended period of time.

James Horne of Loughborough University in England has written extensively on this issue. His position is that people make the error of implementing such a regimen "cold turkey." Sleep is abruptly reduced from the usual 8 or so hours per night to about 6 hours. This leads to daytime sleepiness and decreased efficiency. Horne suggests that total sleep comprises both core and optional sleep periods. Core sleep is the sleep we need; optional sleep we can do without. Core sleep is stages 3 and 4, which occur during the first half of a normal night's sleep (see the hypnogram in Figure 2.2) and, to a lesser extent, REM sleep. Stage 2 makes up the remainder of non-REM sleep and occurs predominantly during the last half of the night. Horne describes findings showing that "a few weeks of careful adaptation will allow a sizable portion of (optional) sleep to be removed without producing increased daytime sleepiness or other ill effects" (Horne, 1988, p. 180).

There are, however, other sleep workers who maintain that the standard 7 to 8 hours of sleep per night for adults is optimal and reduced amounts are detrimental—even when the reductions are gradual over an extended period (Friedman et al., 1977; Mullaney et al., 1977). A final observation on this issue is that each species has a well-defined circadian rhythm governing sleep and waking that has been honed over a long period of evolutionary time. Sleep duration for the human species is clearly in the vicinity of 7 to 8 hours during the dark period. Except for short-term pressures like term papers and exams, it might be prudent to not tamper with this. This conclusion has been reinforced and refined by Kripke (2002) who conducted a large-scale epidemiological study. In 1982 more than 1 million men and women from 30 to 102 years of age were surveyed about the number of hours they slept per night. By 1988 it was found that those who slept 7 hours had the highest rate of survival and those who reported more than 8.5 hours of sleep or less than 3.5 or 4 to 5 hours of sleep had an increased risk of dying.

Short-term sleep deprivation includes total sleep deprivation for one night (an "all-nighter," as students call it) and more unusual, for two or three nights. The urge to sleep, especially during the predawn hours, is overwhelming and requires a good deal of discipline, if not help from a human or a chemical "waker-upper," to avoid lapsing into sleep. In the morning, alertness and performance are unimpaired. Even later in the day, when tiredness manifests itself, subjects can mobilize their resources, rise to the occasion, focus on the task at hand, and perform without impairment. However, if the work or test situation is low key, attention falters and, more seriously, lapses in consciousness (called "microsleeps") occur, and performance suffers. For drivers, the moral is obvious: two to three seconds of a microsleep episode can be fatal.

Experiments using short-term sleep deprivation have been conducted to investigate one of the more widely known hypotheses about the function of sleep, namely, that sleep is important for the storage of memory traces. Since sleep deprivation is just one of the ways this issue has been studied, and since sleep and memory is a large topic in its own right, a separate section in this chapter will be devoted to sleep and memory.

We turn now to long-term sleep deprivation. The ultimate approach to exploring the need for sleep and the physiological and behavioral effects of sleep deprivation has been to push limits by imposing extensive periods of sleep deprivation. There are a number of accounts of self-imposed total sleep deprivation for extensive periods of time. These have captured the imagination of the popular press and of the public. One well-documented case is that of Randy Gardner, a 17-year-old high school student in San Diego, California. As a science fair project, with the help of two friends, he decided to break into the *Guinness Book of World Records* by staying awake longer than the listed record of 260 hours. His goal was to remain awake for 264 hours (11 days). In 1965 Dement, at Stanford University in Palo Alto, California, read about the project in a local newspaper. As a sleep researcher interested in sleep deprivation, Dement volunteered to monitor Randy and observe the effects of prolonged sleep deprivation.

Randy's parents were pleased to have a physician oversee their son for this unusual project. In two interesting accounts of this event, Dement described how difficult it was for Randy and for those who stayed awake with the youth to keep him from lapsing into sleep (Dement, 1976; Dement and Vaughan, 1999). Since very little was known about prolonged sleep deprivation, Dement watched for any physical or psychological changes. He noted that between 3 A.M. and 7 A.M., when it was difficult to stay awake, Randy would sometimes get angry upon being prodded and shaken to prevent lapsing into sleep. At such times physical activity was helpful, and Dement would take Randy outside to play basketball in the backyard. They also visited an arcade and competed against each other on a mechanical baseball machine—with Randy usually winning. Dement noted that Randy's motor coordination and motor skills in general did not suffer from sleep deprivation. Dement commented that "Except for a few illusions—one or two minor hallucinatory experiences—Randy demonstrated no psychotic behavior during the entire vigil, no paranoid behavior, no serious emotional change" (Dement, 1976, p. 12).

At the termination of this endurance marathon after 264 hours, a press conference was held at Randy's home. In attendance were reporters and cameramen from all over the world. As Dement states in his account of the press conference, "At a lectern bristling with microphones, Randy seemed like the President of the United States. He conducted himself flawlessly, neither slurring nor stumbling over words" (Dement and Vaughan, 1999, p. 245). The scientific point is that after 11 days without sleep, Randy was able to rise to the challenge and perform well. There was great speculation about how long Randy would then sleep and if there would be any aftereffects of the prolonged sleep deprivation. Dement stated that "Randy Gardner went to sleep . . . and slept for only fourteen hours and forty minutes, and when he awoke he was essentially recovered" (Dement, 1976, p. 12). Follow-up observations in ensuing years indicated that Randy appeared "to be completely healthy and unaffected by the experience" (Dement, 1976, p. 12).

In view of more recent findings on the effects of extensive sleep deprivation, I expect that sleep experts would strongly discourage any individual who proposed to push the limits of sleep deprivation. Our current knowledge that extensive sleep deprivation may be harmful is based on the dramatic findings of Allan Rechtschaffen at the University of Chicago. He and collaborators deprived rats of sleep and reported their findings in a series of papers starting in 1983 (Rechtschaffen et al., 1983). Much of that work was summarized and elaborated upon in 1989 (Rechtschaffen et al., 1989).

Rechtschaffen and his colleagues (Bergmann et al., 1989) devised an apparatus and procedure for sleep deprivation consisting of a circular platform (a disk) that could rotate above a reservoir of shallow water. The platform was divided in half by a fixed barrier that did not rotate with the disk; that is, the disk rotated above the water and below the fixed barrier. Two rats were placed on the disk, one on each side of the barrier. Both rats had implanted electrodes that monitored the EEG and EMG recordings for sleep and waking. When the rat designated as the experimental animal showed the first sign of sleep the disk rotated, and if the rat remained asleep it would very soon reach the barrier and be deposited into the water. Rats learned to avoid this by awakening just after the platform started moving. The other rat on the opposite side of the wall had no control over the movement of the disk. This rat was designated the yoked control because what happened to it, with respect to the platform rotating and being forced to walk, was yoked (coupled) to conditions imposed on the experimental animal. The control rat experienced the exact same conditions of disk movement as the experimental rat, but the disk movement was not coupled to whether the control rat was awake or asleep. The control rat could peacefully sleep while the experimental rat was awake and the disk not moving. Under those conditions the experimental animals were deprived almost totally of sleep (both REM and non-REM sleep), and the control animals experienced minimal sleep deprivation.

The effects of sleep deprivation were dramatic and extreme. Within two to three weeks every sleep-deprived rat died. Histological examination of the brain and other organs revealed no pathological changes that would cause death. In 1999 Rechtschaffen, in collaboration with J. M. Siegel at UCLA, again used the disk-over-water technique to deprive rats of sleep. In this work, sensitive histological stains were used to detect subtle changes in brain tissue. The stained brain sections suggested that degenerative changes found in areas of the hypothalamus and cerebral cortex may have been responsible for some of the effects of prolonged sleep deprivation (Eiland et al., 1999; Ramanathan et al., 1999).

The prolonged sleep deprivation produced a number of effects that were not lethal per se. The experimental animals were scrawny and had discolored and disheveled coats and skin lesions on their paws and tails. They all showed significant weight loss despite increased food intake (80–100% greater than normal). This last finding was a clue to what caused the biological effects of sleep deprivation and provided a hint to one of the functions of sleep.

Weight loss in the face of increased eating could signify increased metabolism. A cause of increased metabolism when physical work is not driving it is the regulation of body temperature to the challenge of a cold environment. All the sleep-deprived rats showed hypothermia. In fact, if given the choice of an ambient temperature, these rats chose 50°C (122°F)! That temperature is aversively hot for a normal rat or a human. Even if the sleep-deprived animals were kept warm, they still died. If they were removed from the sleep deprivation apparatus just prior to succumbing and permitted to sleep, some died, but others recovered completely after two weeks.

An association between sleep and body temperature has been known for a long time. The circadian rhythms of body temperature and sleep are closely tied; temperature falls at night when one sleeps and slowly rises close to the time of awakening. Also, sleep usually accompanies feverish temperatures like 40°C (104°F). Finally, the neural control of both circadian rhythms (body temperature and sleep) involves hypothalamic and preoptic regions of the brain.

In research during the 1980s, Rechtschaffen and colleagues (Kushida et al., 1989) used the same apparatus to selectively deprive rats of REM sleep. Essentially, the same dire outcome resulted—with the exception that these rats survived about two weeks longer than the rats who were deprived of total sleep (Rechtschaffen et al., 1989). When close to death, some rats were removed from the apparatus and permitted to sleep without interruption. As with Dement's REM-deprived human subjects and my REM-deprived cats, these rats went immediately into REM sleep and showed a large rebound of REM sleep—as much as 10 times more than the normal amount—and recovered from their debilitated state.

Since total sleep and selective REM sleep deprivation produce essentially the same lethal effect, it has been argued that at least the REM sleep component of sleep is necessary for survival. A cautionary note on this point is required. It took about five weeks of REM sleep deprivation to prove fatal, in contrast to three weeks for the rats deprived of total sleep. Could total amount of sleep loss rather than type of sleep loss be the relevant factor? It would be of great interest to deprive animals selectively of non-REM sleep (stages 2, 3, and 4) to determine its effects. However, the problem is that REM sleep normally occurs only following a period of non-REM sleep, so preventing an animal from having any non-REM sleep at all would also result in the deprivation of REM sleep.

A major question was, What kills the animals? What happens during sleep deprivation that is lethal? An answer to this question was proposed by Carol Everson and Linda Toth (Everson and Toth, 2000). Using the disk-over-water method, Everson and Toth deprived rats of total sleep for 5, 10, 15, and 20 days. After their prescribed days of sleep deprivation, the rats and their yoked controls were killed and tissue samples from various organs were collected. Since earlier evidence showed that sleep-deprived rats sustain serious bacterial infection (Everson, 1993), these tissues were cultured to identify the species of bacteria that were present and perhaps determine the source of the infection.

As found in the earlier work, the sleep-deprived rats consumed more food but lost weight, developed skin lesions, and showed a general deterioration of health. These signs were progressive, and by the twentieth day the rats in the survival group were severely ill. In the previous work, when sleep deprivation was permitted to extend to its lethal culmination, the proximal cause of death was a widespread bacterial infection of the bloodstream with no discernible infectious focus.

To determine the source of the infection, Everson and Toth conducted painstaking analyses of bacterial cultures derived from the spleen, liver, kidney, skin, and intestine (including contents). Also cultured was tissue from lymph nodes located in various regions of the body and swab specimens from the peritoneal, thoracic, and throat cavities. Blood removed directly from the heart was also cultured. Two sources of bacterial infection were found. One was from the external environment that found entry via the skin lesions on the paws and tail. This did not appear to be the cause of serious consequences. The other bacterial source was the intestine. Sleep-deprived rats had significant overgrowth of bacterial species normally found in the intestine and normally contained within the intestine. This proliferation of intestinal bacteria was associated with a bacterial invasion from the intestine to the lymph nodes adjacent to the intestine (the mesenteric lymph nodes). These lymph nodes ordinarily serve as the first line of defense of the immune system against bacteria that leave the intestine and enter the internal environment. (The gastrointestinal tract from mouth to anus is considered to be part of the external environment. Its contents become part of the internal environment only after passing through the gut wall.) The lymphatic component of the immune system appears to be severely compromised in sleep-deprived animals. As a result, not only the lymph nodes, but also other organs were colonized by various species of bacteria. Eventually, the buildup of bacteria in the internal organs and lymphatic system resulted in infection of the bloodstream and shortly thereafter, death.

These findings indicate that sleep deprivation compromises, if not disables, the immune system at the level of local lymph nodes as well as at the systemic level. That's serious. The data also show (using the 5-day deprivation group) that this process starts at least as early as the fifth day of sleep deprivation. The minimum period of sleep deprivation needed to produce an impairment of the immune system is not known. Could reduced sleep for just a few days impact the immune system? Everson and Toth end their paper with the suggestion that sleep-deprived individuals are more susceptible to disease and that already sick patients who are sleep deprived would become more severely ill and their recovery would be impaired. (A serious problem in hospitals is the number of patients who contract bacterial infections during their hospital stay. Aside from the standard precautions, perhaps the hospital routines that involve sleep interruptions should be avoided.)

The next question is, Why does sleep deprivation compromise the immune system? Or, put another way, What is it about sleep that maintains the integrity of the immune system? This is clearly an important area to be researched.

It should be pointed out that Rechtschaffen does not agree with the conclusion of Everson and Toth that a breakdown of the immune system was the cause of death. In a letter to the editor of the journal in which the Everson and Toth paper appeared, Rechtschaffen and Bergmann (2001) refer to a 1996 paper in which they and their colleagues (Bergmann et al., 1996) reported that antibiotic treatment to eliminate infection did not protect sleep-deprived rats from dying. As is often the case in scientific controversy, Everson and Toth responded with a letter that defended their position. The issues are quite technical and complex and the controversy is by no means resolved.

In addition to the above controversy about total sleep deprivation, the finding by Rechtschsaffen and colleagues (Kushida et al., 1989) that selective REM sleep deprivation in rats was lethal is in conflict with the few reports in the literature of individuals who have minuscule, if any, amounts of REM sleep and show no obvious signs of impairment. One documented case, described by Peretz Lavie (1986), is of an Israeli war veteran who had received shrapnel wounds to the brain. All-night sleep recordings for 8 nights revealed that on most nights there were no traces of REM sleep at all; on a few nights there were minimal traces that amounted to 2 to 5% of total sleep time. When computerized brain imaging was used to determine the locations of shrapnel wounds, a sliver was found in the region of the pontine brain stem known, from the animal literature, to control the occurrence of REM sleep. Except for motor and speech impairment related to cerebellar and cerebral damage, there were no other deficits. Memory and other cognitive processes were intact.

A similar observation that is inconsistent with the dire effects of REM sleep deprivation in rats comes from the pharmacological treatment of clinically depressed patients with two types of antidepressant drug, the monoamine oxidase (MAO) inhibitors and the tricyclic antidepressants. Patients who require large doses of these drugs over extended periods of time to control depression also show decreased to almost no signs of REM sleep. Yet these patients show no negative reactions to the loss of REM sleep. In fact, their loss of REM sleep may be responsible for the antidepressant effect. In 1975 Gerald Vogel and his coworkers reported that depressed patients who were deprived of REM sleep experienced alleviation of their depression (Vogel et al., 1980). This finding has been replicated and has since been used as a nonpharmacological treatment for depression with no reports of adverse effects on cognitive or memory processes.

Another effective treatment for clinical depression is electroconvulsive shock (ECS). Is there a relationship between ECS and REM sleep deprivation? This was shown to be the case in a series of studies between 1965 and 1970 by Cohen and Dement. They reported that ECS in rats and cats reduced the amount of REM sleep and eliminated the strong REM sleep rebound after REM sleep deprivation. They also showed that after REM sleep deprivation, the amount of electrical current required to produce a convulsion was decreased; that is, REM sleep deprivation decreased the threshold for ECS (Cohen and Dement, 1965; Cohen et al., 1967, 1970). Apparently, depressed patients can benefit from a treatment that increases neuronal excitability. ECS certainly does that in a massive and dramatic way, and REM sleep deprivation also appears to do that: the threshold for ECS decreases, and depression is reversed. There is certainly an interesting three-way relationship between REM sleep deprivation, ECS, and depression that could be further explored to shed light on the possible function of REM sleep in affecting neuronal excitability.

A final issue to consider about the lethal effect of sleep deprivation in the rat is the possible stressfulness of the deprivation procedure.

Rechtschaffen's controls for this were good. Both experimental and yoked controls experienced the same conditions except for the increased number of awakenings experienced by the experimental rats. Along with the sleep deprivation (total or REM), repeated awakenings might be stressful. In Rechtschaffen's early work in the 1980s, postmortem histology revealed no differences between organ tissues removed from experimental and yoked control rats. Intestinal ulcers that often accompany stress, were not observed. Hormones related to stress, ACTH from the pituitary gland and cortisol from the adrenal cortex, were elevated in both experimental and control animals. So both groups, not surprisingly, by that indicator, were somewhat stressed. However, in the days just prior to death, cortisol levels in the sleep-deprived animals rose significantly. Again, it is not surprising that a manipulation potent enough to cause rampant bacterial infection, plus the infection itself, produced increased stress. It is not known whether the loss of sleep independent of the method imposed to prevent sleep would be lethal. Does sleep loss per se compromise the immune system, or does the stress associated with an imposed regimen of sleep deprivation impair the immune system?

Consider the animal work of Rechtschaffen and of Everson and Toth versus the human cases in which sleep loss caused by antidepressants and by brain lesion was not lethal. Could it be that the forced deprivation of sleep by an imposed stressful situation carries consequences that, in the short run, are detrimental and, in the long run, are lethal? Experimental findings indicate that the lethal outcome of forced sleep deprivation is due to the malfunction of two physiological, regulatory systems: there is impairment in the hypothalamic mechanism that regulates body temperature, and the immune system that protects against infection. It may be that the imposed loss of sleep and its concomitant stress interferes with the integrity of two critical physiological systems.

Sleep and Memory

A view held by a number of sleep researchers is that the effective storage of memory traces, often referred to as "memory consolidation," is facilitated by sleep. The most prominent variant of this position is that during REM sleep, the neural circuits involved in the learning and storage of recent memories are reactivated and some form of rehearsal* oc-

---

*The term "rehearsal" implies a conscious, cognitive process; however, the memory trace may be strengthened by a simple mechanistic reactivation of the neural activity or by biochemical events involved in the learning experience, without the involvement of a cognitive component.

curs to strengthen the memory trace. Recently, sleep researchers have paid attention to advances in cognitive science showing that there are different types of learning and memory systems and that sleep is important for memory consolidation. The early work focused mainly on REM sleep and will be reviewed first, followed by the recent and interesting findings on the role of both types of sleep in memory.

There is substantial evidence that the neural processes involved in the storage of memory traces are labile for some period of time after a learning experience and can be interfered with (as by electroconvulsive shock) or enhanced (as by practice). Related to this is the notion that if subjects are deprived of REM sleep (and of rehearsal) after a learning session, memory is impaired. This is in contrast to subjects who are permitted to have REM sleep between the learning and testing periods and are thus able to rehearse the learning experience. This view holds that rehearsal occurs during REM sleep and serves to facilitate the consolidation of the labile memory trace, that is, to convert it into a more stable trace.

Research investigating REM sleep and memory has been conducted in animals and in humans. Two major strategies have been used. The first is the obvious one: prevent subjects from having REM sleep for a period of time following learning trials and test to see whether memory for the task has been impaired. These subjects would be compared with those who are permitted to sleep normally after the original learning or who are deprived of non-REM sleep for an equivalent period of time. The other strategy was to present a learning session and see whether the sleep period following the learning experience showed a greater amount of REM sleep than that of subjects who did not have a learning session prior to their sleep. The rationale here is that during the REM sleep following learning, something more is happening, perhaps memory traces of the learning experience are reactivated (rehearsed), and this process results in an increased amount of time spent in REM sleep. This is in contrast to the amount of REM sleep seen during the sleep following a waking period in which there is little if any learning.

As one would imagine, when the REM sleep deprivation approach is used, there must be control conditions to permit confidence that an effect, if seen, can be attributed to the loss of REM sleep rather than to some other cause. An example of this that has plagued this area of research is that the method used to prevent REM sleep imposes some stress on the subjects. The problem, of course, is to know whether the deficit in retention is due to the REM sleep deprivation itself, to the stress, or to some other factor associated with the procedure to prevent REM sleep. Another and somewhat more subtle issue is the interpretation of positive findings—even if one is satisfied that the learning deficit is due to REM sleep deprivation. Could REM sleep deprivation cause a learning deficit for reasons other than interference with a memory consolidation process?

An example of this possibility was provided by Gerald Vogel, whose work was discussed earlier in this chapter in a related context. Vogel contends that REM sleep deprivation increases drive and arousal levels (Vogel, 1979). He and others point to evidence that after REM sleep deprivation, the heightened arousal is accompanied by increased sensitivity to stimuli that have distracting and interfering effects during the original learning and during tests for retention. This is especially the case with difficult learning tasks (McGrath and Cohen, 1978; Horne, 1988). Albert et al. (1970) presented evidence for this position. They reported that REM sleep deprivation in rats did not interfere with retention of avoidance learning but did produce increased levels of activity and increased sensitivity to sensory stimuli. They concluded that when REM sleep deprivation interferes with learning, it does so not because of interference with the memory trace, but because REM sleep deprivation increases excitability and distraction, and that is what interferes with learning and memory consolidation.

In addition to the above observations that cast doubt on a relationship between REM sleep and memory are other findings that do not support such a relationship. In the preceding section on REM sleep deprivation, we described observations that losses of REM sleep due to the use of antidepressant drugs or to a brain stem lesion were not accompanied by memory deficits. Despite the complications for the position that REM sleep is important for memory, some researchers are confident that REM sleep deprivation does interfere with the memory consolidation process (Fishbein and Gutwein, 1977, 1981). Other workers suspend judgment on results arrived at by this approach and place more credibility on findings that use an experimental strategy that does not prevent REM sleep. This strategy is designed to determine whether something of importance for memory consolidation occurs during REM sleep.

This approach is exemplified by research from three laboratories. The work of Elizabeth Hennevin and her colleagues at the University of Paris-South and Carlyle Smith and his colleagues at Trent University in Ontario, Canada, provide strong evidence that memory consolidation occurs during REM sleep periods that follow learning (Hennevin et al., 1995; Smith, 1995, 1996). These investigators use similar methods, and their work will be discussed together. The research of Pierre Maquet at the University of Liège, Belgium, and University College, London, uses brain imaging techniques and will be described separately at the end of this section. The study of REM sleep following learning has been conducted with animal and human subjects using a variety of learning tasks. The animal research of the Hennevin and Smith groups will be described first. Since they both tend to use the term "paradoxical sleep" rather than "REM sleep" in their rat experiments, I will do so too in describing their work.

In learning trials distributed over a number of successive days, performance improvement does not always occur in a smooth incremental fashion, as the classic learning curve is often envisioned. Over a number of days there are small incremental changes of performance and then, on a subsequent day, a sudden large increment (a surge or spurt) of improved performance occurs. Both the French and Canadian research groups observed that during the sleep period just prior to such a large performance increase, there was a marked increase in the amount of paradoxical sleep time. The supposition is that during the paradoxical sleep preceding the improved performance, a considerable amount of memory processing and consolidation had occurred and that, in turn, produced the performance spurt. When performance leveled off, paradoxical sleep time reverted to baseline level. The learning had occurred and no further consolidation was necessary, and the amount of paradoxical sleep was the same as it was prior to the imposed learning. If the task was modified after performance leveled off, paradoxical sleep time again increased in parallel with the improved performance on the modified task.

Other evidence that paradoxical sleep is important for memory consolidation is based on the common observation of individual differences in learning. Even in a genetically homogeneous population, such as a particular strain of laboratory rats, some learn a task rapidly, some learn slowly, and some rats do not learn at all, especially if the task is difficult. Both Hennevin's and Smith's groups reported that rats that learned showed increased paradoxical sleep during their learning days, and rats that did not learn did not show an increase in paradoxical sleep during the course of learning.

A procedure and concept introduced by Smith and his colleagues has yielded interesting findings. They observed a limited time period in the sleep that followed a learning session during which the amount of paradoxical sleep showed a marked increase. For different types of tasks, the limited time of enhanced paradoxical sleep was shown to occur at different time periods past the end of the learning trials. The time period of increased paradoxical sleep was called the paradoxical sleep window. Once the window for a particular task had been determined, rats were selectively deprived of paradoxical sleep during that period of time. These animals suffered memory impairment. Control rats that were deprived of paradoxical sleep during periods outside the window and other controls that were deprived of slow-wave sleep for a like duration of time showed no memory deficit. Smith and coworkers showed not only that paradoxical sleep is important for the process of memory consolidation, but also that whatever is happening is happening only during limited, selective periods of time following the learning experience.

Recent work on sleep and memory uses brain imaging techniques to determine the brain regions that are selectively active during REM sleep. This type of data was discussed briefly in Chapter 12, The Neural Control of REM Sleep. More relevant to the current topic is whether some aspect of that brain activity can serve as an indicator (or marker) for memory processes that occur during REM sleep.

Pierre Maquet used the technique of positron emission tomography (PET) and regional cerebral blood flow (rCBF) measurements to visualize regions of the human brain that are metabolically active during learning (Maquet et al., 2000). Increased cerebral blood flow to a region is reflective of increased metabolism that in turn is an indicator of increased synaptic activity and increased firing of nerve cells in that local area. This is not a trivial procedure; it involves a great deal of patience on the part of subjects and expensive equipment.

Each subject received a venous catheter that provided a slow infusion of radioactive water, $H_2^{15}O$, and was adapted to having his head stabilized in a fixed mask and head holder so the brain could be scanned by the PET apparatus. The PET scans detected changes in regional blood flow during periods of learning trials as well as during sleep. Different groups of subjects were exposed to the same training task but to different sleep conditions. They were all trained on a visual–motor reaction time task. A signal light appeared on a computer screen in one of six fixed positions, and the subject was instructed to press one of six keys on a keyboard that corresponded to the position of the light. The reaction time from the signal light onset to the key press response was measured on each trial. One group of subjects received the PET scan when they were awake and resting and again when they were performing and learning the reaction time task. The two scans from each subject of this group yielded baseline information about the brain areas that are active during acquisition of this task. Other groups were trained during sessions in the afternoon and then scanned while still awake and again during sleep that night. Some subjects were not trained, but were scanned during sleep to determine the regional blood flow pattern during REM sleep when training had *not* been imposed earlier. Other subjects that received training were scanned during a night's sleep and received another training and reaction time session following their sleep.

Maquet and his colleagues found that reaction time performance during the session following sleep showed an improvement. The PET data showed that certain brain areas were metabolically active during performance of the visual–motor learning task. These involved, as one would expect, visual, motor, and somatosensory (the tactile component of the response) areas of the brain. Of greater interest was the finding that some of those regions were also active during REM sleep periods that followed training. This is direct evidence that the brain structures involved in a learning task while awake are, in a sense, reactivated (possibly in a similar fashion) during REM sleep. This finding, in conjunction with the improved performance after a night's sleep, supports the position that during REM sleep following a learning experience there is a neural form of rehearsal and a consolidation of the memory trace.

Further evidence that brain structures involved in learning are reactivated during sleep utilizes brain recordings from behaving rats taken during learning trials and again, from the same electrodes, during sleep. This approach has been particularly successful during place learning, in which an animal learns its way around in an initially strange place. It has been known for a number of years that certain nerve cells in the hippocampus can be described as "place cells." O'Keefe and colleagues showed that specific cells were selectively active and fired at higher rates when the rat entered a certain location. Other hippocampal cells were active when the rat was in a different location. As the rat revisited the same locations on subsequent trials, the same cells displayed an increased rate of firing (O'Keefe and Dostrovsky, 1971; O'Keefe, 1976). Subsequently, Pavlides and Winson (1989) and Wilson and McNaughton (1994) replicated the O'Keefe findings. They further demonstrated that the cells that increased their firing when the animal was in a specific location during a learning experience were similarly active during the slow-wave sleep that followed the learning experience. This neural activity did not occur during sleep that preceded learning trials. Pavlides and Winson reported the increased firing during both slow-wave and REM sleep, but the increase during slow-wave sleep was greater. Wilson and McNaughton reported the increased firing of place cells only during the slow-wave sleep state.

Recent findings in cognitive science reveal the existence of different types of learning and memory systems, and recent work in sleep research suggests that slow-wave and REM sleep are differentially involved in the consolidation of different types of memory. The major distinction is between what cognitive scientists call "procedural" and "declarative" memory. Procedural memory refers to the learning and retention of certain skills, particularly sensory-motor skills. One is not consciously aware that such skills are being laid down in memory. This process is sometimes also called "implicit learning," and its storage is called "implicit memory." Declarative memory, also called "episodic" or "explicit memory," is based on the recall of episodes of which one is consciously aware. A child learning to ride a bicycle provides examples of the two types of memory. The child would be consciously aware of and be able to recall an episode in which he or she fell and was seriously hurt. This is episodic or explicit learning and memory. However, the child does not have a conscious awareness of the vestibular- and proprioceptive-motor skills being laid down in memory during the process of learning to ride the bike. This is procedural or implicit learning and memory.

In light of the past emphasis on REM sleep and memory, it is interesting to note that recent work suggests that slow-wave REM sleep is also important for memory storage. There are indications that the two types of sleep are differentially important for the two types of learning and memory. Two types of sleep, two types of memory? Suggestive? Are the two types of sleep related to the two types of memory? Is X-type sleep important for A-type memory and Y-type sleep important for B-type memory?

Most studies in this area of research use human subjects and capitalize on the observation that slow-wave sleep (stages 3 and 4) dominate the first half of a night's sleep and REM sleep occurs predominantly during the second half of the night. (Review the hypnogram in Chapter 2, Figure 2.2.) Plihal and Born (1997), in Lübeck, Germany, utilized a first half-night, last half-night design. Subjects were trained on a declarative (verbal learning) task and on a procedural (mirror tracing) task during the day and then tested after being permitted to sleep either during the first half or last half of the night. Subjects trained on the declarative task and permitted to sleep through the first half-night (dominated by slow-wave sleep) performed better than control subjects who remained awake after the learning experience. Having slow-wave sleep, however, did not help in remembering the procedural motor task. In contrast, the procedural skill was performed better after being permitted to have REM sleep (during the last half of the night), compared with subjects who did not have their late-night REM sleep. Having REM sleep but little slow-wave sleep did not facilitate performance on the declarative task.

A second study by these authors (Plihal and Born, 1999), using different procedural and declarative tasks, confirmed their earlier work (Plihal and Born, 1997). These studies appear to support the simple X-A, Y-B relationship just alluded to: slow-wave sleep is important for declarative memory; REM sleep is important for procedural memory. However, the same group reported an experiment that complicates the story considerably, but in ways that bring us closer to recognizing the complexities and subtleties of how sleep is involved in learning and memory.

Gais et al. (2000) used a visual discrimination task that they and earlier workers (Karni and Sagi, 1991; Karni et al., 1994) describe as a procedural memory task. The task was to report whether lines briefly flashed on a computer screen were oriented in a horizontal or vertical array. After all subjects had received a training session during the day, one group was permitted to sleep only during the first half of the night, another group slept only during the second half of the night, and a third group was permitted to sleep the entire night. Performance on this procedural task did *not* improve after late-night sleep (abundant in REM sleep) was permitted and did improve after early-night (slow-wave) sleep was permitted. These findings contradict their two earlier reports (Plihal and Born, 1997, 1999). Interestingly, however, the third group of subjects, who were permitted to sleep the entire night showed, more than three times greater improvement over the improved performance of the early-night (slow-wave) sleep subjects.

Two main conclusions were drawn from these data. One is that memory consolidation occurs during the night as a two-step process. The REM-sleep-related step occurs predominantly during the second half of the night's sleep and is mainly effective after memory processes have already been initiated during the early hours of sleep when the step related to slow-wave sleep occurs. A second conclusion is related to the conflicting findings with their earlier data. The authors suggest that the procedural tasks used in the two earlier experiments were more complex than the task of perceiving line orientation used in their year

2000 study, and that "REM sleep may become increasingly important with more complex tasks" (Gais et al., 2000, p. 1338). The notion here is that memory consolidation for complex tasks requires the occurrence of REM sleep, and in the earlier studies, which had used complex procedural tasks, improvement was shown when REM sleep was permitted. In the study that used the relatively simple perceptual task (Gais et al., 2000), the role of REM sleep was demonstrated only after a consolidation process had already occurred earlier in the night.

Recent additional support for the position that memory consolidation occurs during sleep and involves a two-step process is provided by Stickgold et al. (2000 a, b) for the human and by Louie and Wilson (2001) for the rat. Stickgold and coworkers, using the same visual discrimination task developed by Karni and Sagi (1991) and used by Gais et al. (2000), reported that "sleep within 30 hours of training is absolutely required for improved performance" (Stickgold et al., 2000a, p. 1237). Subjects trained and tested later that day but before any sleep had occurred showed no improvement. Subjects kept awake the first night after learning and permitted to sleep two full nights later also showed no improvement. The only subjects that demonstrated improved performance had been permitted to sleep during the first night after the day of training. In another experiment, Stickgold and colleagues (2000b), using the same visual discrimination task, showed that the memory consolidation during the first night of sleep following a learning experience occurs as a two-step process. Their data analysis showed that performance on the task increased on the day following a night's sleep in proportion to the amount of REM sleep during the last quartile of sleep (the last 2 hours of an 8-hour sleep period) and also was proportional to the amount of slow-wave sleep during the first quartile of sleep (the first 2 hours of sleep). They provide evidence that both slow-wave sleep in the early part of the night and REM sleep near the morning "appear to be required for overnight learning" (Stickgold et al., 2000b, p. 250).

The rat experiments by Louie and Wilson (2001) complement the human findings about a two-step process in memory consolidation during sleep. They report that during the REM sleep that follows a place-learning experience, hippocampal cells fired at a higher rate. The increased firing rates occurred over durations of tens of seconds to even minutes during the learning and during the reproduced neural activity in subsequent REM episodes. In contrast, the reactivation of place cells during slow-wave sleep that had also been observed by Wilson and

McNaughton (1994) occurred during a short time period, on the order of seconds. The work of Louie and Wilson provides evidence from rats that both REM and slow-wave sleep are involved in memory.

Evidence from the animal physiological and human neurological literature on memory indicates that memory traces are stored initially in neural circuits of the hippocampus, and more permanent long-term storage occurs at the level of the cerebral neocortex. Louie and Wilson suggest that during the place-learning experience, new information activates hippocampal cells for an initial processing of memory. During subsequent slow-wave sleep, hippocampal discharges are reactivated to establish hippocampal–cortical circuits for a first stage of memory consolidation during sleep. During later REM sleep, the cortical circuits engage the hippocampus to reactivate the neural trace of the learning experience. This cortical–hippocampal network "replays" the memory trace for durations as long as minutes and comprises the second stage of memory consolidation during sleep. Stickgold et al. (2000b) posit a similar sequence of events for memory consolidation that occurs during slow-wave and REM sleep.

Current research on sleep and memory has incorporated the concepts and experimental paradigms of memory research being developed in cognitive science. This is a relatively new field, and some confusion and contradictions exist about which stages of sleep are important for the consolidation of which types of learning and memory. I expect this situation to continue until the distinctions between declarative, procedural, explicit, and implicit learning and memory—and categories yet to be described—are sorted out. The distinction between declarative and procedural may be too simple. Factors such as degree of complexity and of cognitive and affective involvement might also be important to consider. As Carlyle Smith points out, "the term procedural may not be broad enough to include all the types of learning that are not declarative . . . (and) procedural motor memory may not be a unitary system but can be dissociated depending upon the amount of cognitive involvement" (Smith, 2001, p. 501).

The recent work just described suggests that enhanced memory consolidation is due to a reactivation or replaying of neural memory traces during sleep. Frank et al. (2001) present another possibility. They found that experience-dependent neural plasticity in the developing visual cortex of the kitten brain was enhanced during non-REM sleep. They suggest that alternatively, or in addition to a neuronal replay hypothesis, biochemical events known to increase during sleep may strengthen the memory trace. They refer to the release of growth factors and increased protein synthesis during slow-wave sleep that could enhance neural plasticity and memory traces during sleep. This is a more mechanistic, less cognitive process than that implied by the notion of replaying memory traces. Frank et al. cite the research of a number of investigators to support this possibility (Ramm and Smith, 1990; Nakanishi et al., 1997; Cauter and Spiegel, 1999).

As with the previous topic, the effects of long-term sleep deprivation, the research on sleep and memory also has its controversies. For a reasoned critique of the REM sleep-memory consolidation hypothesis see J.M. Siegel's essay in Science, 2001.

14

Disorders of Sleep and Waking

Wilse Webb, emeritus professor of psychology at the University of Florida and a pioneer in sleep research, distinguished between sleep disturbances and sleep disorders. The former are "disturbances of normal and healthy sleep patterns that may or may not impact waking performance. . . . Sleep disorders (are) clinically diagnosable conditions of the sleep/waking system that require therapeutic intervention" (Webb, 1992, p. 80).

Webb describes a study showing that normal subjects would sleep an hour or so longer in the morning if not for the alarm clock. When permitted to do so, they were more alert and did not experience periods of sleepiness during the day (Webb, 1992, p. 83). Why the chronic need for sleep? Since the advent of electricity and the lightbulb, we have more hours in the 24-hour day in which to be active. In our modern society, we have found more things to do to fill those artificially illuminated hours (albeit, some of that time is spent in the dark—in front of an electrically lit cathode ray tube, the television set). Under these circumstances many of us experience some degree of sleep deprivation. However, we often have the opportunity to sleep a little longer and manage to function, perhaps not at peak efficiency, but reasonably well. In some circumstances, however, sleep deprivation poses very serious societal problems. For example, long-distance truck drivers and shift workers who are chronically sleep deprived are prone to lapses in attention, especially under conditions of monotony, that can be disastrous to themselves and others. Sleep-deprived subjects may experience microsleep periods lasting between 5 and 10 seconds to up to a minute. For a person at the wheel of a fast-moving vehicle or in an industrial setting, these lapses can be lethal.

Even though some sleep disturbances may not require clinical diagnosis, they may indeed be serious problems and call for some sort of intervention. Such disturbances include those that occur with aging. As we grow older, we tend to have more complaints about the quality of sleep. Typical findings on sleep and aging show that about 5% of young adults are "poor sleepers" in contrast to over 20% of 60-year-olds. The complaints often are related to not getting enough sleep because of difficulty falling asleep, disruptions of sleep, or awakening too early. When such disturbances are mild, we usually cope with them with minimal harmful effects. However, when sleep disturbances cause serious problems, they should be considered to be disorders that require treatment.

Excessive daytime sleepiness is a frequent complaint of patients seen in sleep clinics and is a major cause of societal problems associated with sleep/waking disorders. These problems range from relatively innocuous ones like nodding off during a concert or a lecture to nodding off while driving a car. Motor vehicle accidents due to sleep-deprived drivers falling asleep at the wheel are but one example of sleep-related accidents. According to the National Sleep Foundation, there are 100,000 such accidents per year in the United States resulting in 1500 fatalities. If industrial accidents due to sleepiness-induced lapses of vigilance are included, the human life and monetary costs are considerably higher. A dramatic example is the explosion of a nuclear reactor in Chernobyl, Ukraine, that occurred on April 26, 1986. An investigation indicated that a sleep-deprived crew working at 1:30 A.M. failed to notice that the temperature in one of the reactors had reached a dangerously high level. When they did notice the problem, their impaired reactions were wrong and the nuclear reactor blew. People at the site of the explosion were immediately killed, and many more died soon after from radiation poisoning and burns. The spread of radioactive debris covered an area of more than 2000 square miles, with the result that 17 million people suffered some degree of radioactive contamination. In addition, water and food supplies were contaminated. Those were the acute effects at the time of the explosion. The long-term effects, such as the increased incidence of cancer, birth defects, and immune system deficiency (called by the locals "Chernobyl AIDS") are still being assessed.

In the 1970s, clinical workers in sleep research who were members of the Association for the Psychophysiological Study of Sleep (APSS) recognized a need to categorize and formally describe the many sleep disorders they observed. Their model was the classification and description of medical and psychiatric disorders. In 1979 the American Sleep Disorders Association (ASDA), an offshoot of the APSS, generated a classification scheme of sleep disorders called the Diagnostic Classification of Sleep and Arousal Disorders (DCSAD). In 1990, the ASDA, in collaboration with sleep societies in Europe, Japan, and Latin America, modified the 1979 scheme to produce the International Classification of Sleep Disorders (ICSD) and a manual for the diagnosis and coding of sleep disorders. This classification scheme, revised in 1997, is described in a chapter by Michael Thorpy in the *Principles and Practice of Sleep Medicine,* edited by Kryger, Roth, and Dement (Thorpy, 2000). This third edition of a 1335-page volume is rich with basic scientific and clinical information on sleep disorders. The first edition was published in 1989. I point this out to underscore the puzzling fact that until 1996, the American Medical Association did not recognize sleep medicine as a subspecialty.

The ICSD lists four categories of sleep disorders.

1. The **dyssomnias** include difficulties in initiating and maintaining sleep. This disorder is ordinarily called insomnia, the major cause of excessive daytime sleepiness. Subsumed under this broad category are narcolepsy, sleep apnea, and circadian rhythm disorders such as jet lag and shift work sleep disturbance.
2. The **parasomnias** are disorders of arousal, REM sleep, and the transition state between sleep and waking. Included in this category are sleepwalking, sleep terrors, sleep talking, sleep paralysis, nightmares, sleep enuresis (bed-wetting), sleep bruxism (teeth grinding), and excessive snoring. Included here too are sudden unexplained nocturnal deaths of adults, sudden infant death syndrome (SIDS), and infant sleep apnea.
3. The **medical or psychiatric disorders** are sleep problems associated with mental disorders such as the psychoses, depression, and anxiety, and neurological disorders such as Parkinson's disease, Alzheimer's disease, and epilepsy.
4. The **proposed sleep disorders** are newly recognized sleep disorders on which limited information exists. These include short and long sleep times, menstrual- and pregnancy-associated sleep disorders, and night sweats.

With respect to medical disorders, disorders of sleep and waking are not as serious as diseases such as cancer and heart or liver failure in terms of physical pain and morbidity. Neurological and psychiatric disorders like Alzheimer's disease, schizophrenia, and depression are also not accompanied by physical pain, but their severity and resulting debilitation are undeniable. However, as pointed out by workers in sleep medicine, the costs for the individual and for society of sleep/waking disorders are considerable.

I will describe three sleep disorders that are responsible for major impairments and disruptions in the lives of people. These are insomnia, sudden infant death syndrome (SIDS), and narcolepsy. Insomnia and its main consequence, excessive daytime sleepiness, is by far the most frequently reported sleep disorder. There are multiple causes of insomnia that cut across the four categories of the sleep disorders; I will describe insomnia caused by sleep apnea. Sudden infant death syndrome is a tragic disorder that claims the lives of infants less than a year old. Narcolepsy is a dramatic syndrome often referred to as a disorder of waking. Recent advances in understanding the neurological and genetic basis of this disorder will be described.

Insomnia and Sleep Apnea

Apnea is a temporary cessation of breathing that lasts 10 seconds or more. When it occurs frequently during sleep, it is a major of cause of insomnia. Each episode of sleep apnea causes a brief arousal, which is fortunate, since breathing resumes at this point. However, if cycles of sleep apnea and arousal are frequent, there is a great deal of sleep disturbance, resulting in daytime sleepiness. If the apnea is prolonged, its consequence is death. What causes sleep apnea? Why do people suddenly stop breathing—and why does this happen mainly during sleep? There are two types of apnea, central and obstructive apnea.

Central apnea is due to a malfunction of receptors and associated neural circuits specialized to detect changes in the concentrations of oxygen (O_2) and carbon dioxide (CO_2) in the blood. When respiration slows (hypopnea) or, more drastically, ceases, the amount of O_2 in the blood decreases and the CO_2 level goes up. This condition is the stimulus for these blood gas sensors to excite their target neurons in the respiratory center of the medulla. The inspiratory part of this neural center controls the muscles of respiration to increase the rate and depth of inspirations and thus bring in more O_2. Similarly, increases in the expiratory phase results in more CO_2 being expelled. This corrects the condition that activated the O_2 and CO_2 chemoreceptors, and respiration will return to its basal level. Normally, these cells are sensitive to small changes in blood gas concentrations and small corrective changes in breathing rate will occur, or we will occasionally take a deep breath. These sensors usually work perfectly well during waking and sleep. However, in some individuals the sensors are less sensitive during sleep, and a greater O_2 deficit and a greater CO_2 concentration are needed to activate them. These individuals will briefly arouse, breath more deeply and frequently for a short period, correct their blood gas levels, and fall asleep again—until the next apneic episode. In some cases, these episodes will occur several hundred times during a night's sleep, and usually the individuals are unaware of these multiple sleep interruptions! Nevertheless, they have had an inordinate number of awakenings, and the serious sleep debt thus incurred results in excessive daytime sleepiness—which the apneic persons cannot explain.

Aside from daytime sleepiness, and its consequences, serious organic medical problems have been attributed to sleep apnea. Long-term effects of apnea produce cardiovascular disorders such as arterial damage and chronic high blood pressure—both of which increase the risk of heart attack and stroke. Sleep apnea is now recognized as a major health hazard, and considerable attention is devoted to its treatment.

Obstructive sleep apnea, as the name implies, is due to an obstruction in the upper airway leading to the trachea and lungs that reduces airflow and is signaled by increased respiratory effort. The most frequently used noninvasive treatment for this condition is a device, essentially an air pump, that provides a continuous positive airway pressure (CPAP) to the sleeping person. This machine delivers a constant gentle stream of air to a mask worn over the nose and mouth or over the nose alone. The positive pressure is adjusted so that it is just adequate to keep the airway open when it usually closes during sleep in apnea patients. This treatment has also been used for those who suffer from central sleep apnea. People who adjust to sleeping with this mask report that they awake rested and no longer experience debilitating daytime sleepiness. However, for those who do not tolerate the mask, other options are available.

Another noninvasive procedure, often applied by dentists, uses a mandibular advancement appliance. The mandible is the bone in the lower jaw that carries the lower teeth and soft tissue of the lower or bottom parts of the mouth and throat. The maxillary bone carries the upper teeth and top parts of the oral cavity. The openings for the esophagus and trachea are located where the soft tissues of the upper and lower jaw meet at the back of the throat. This area at the back of the oral cavity is called the pharynx. The esophagus carries solid food and liquid to the stomach; the trachea carries air into and out of the lungs. In some people the lower jaw is somewhat tucked back and the soft tissue near of the opening to the trachea will tend to fold over and partially obstruct the opening. This is ordinarily not a problem during waking. But, especially for overweight people with large necks, the excessive soft fatty tissue and the enlarged base of the tongue are relaxed during sleep, and the opening to the trachea is in danger of being closed over. This is particularly the case when the person is sleeping on his or her back. Snoring occurs under these circumstances because the excessive soft tissue at the opening vibrates vigorously on the inspiration and expiration phases of breathing. In some cases, the loud snoring will abruptly stop. One would think that for a bed partner this would be a relief. However, those bed partners who are awake learn to recognize that the sudden cessation of snoring also signals the cessation of breathing. The soft tissue has, at that point, closed the opening to the trachea. This, of course, is an episode of obstructive sleep apnea. The apneic episode will continue for seconds and even up to 2 minutes until the person briefly arouses, usually with a sudden loud snort and large inspiration. If these sleep interruptions happens frequently during the night, excessive daytime sleepiness results.

Patients with these symptoms have the option of a mandibular appliance that is attached to the lower jaw. This device gently but firmly pulls the lower jaw forward to the extent that soft tissue of the mandible at the back of the throat will no longer occlude the opening to the airway. This device, like other clinical procedures, does not always work, and some patients cannot tolerate the discomfort. To improve the effectiveness and decrease the discomfort, dentists and others have devised over 40 oral appliances for the treatment of obstructive sleep apnea.

For individuals who do not tolerate the CPAP mask and prefer not to use a dental appliance, an invasive alternative may be taken: surgical removal of portions of excessive soft fatty tissue or soft muscle tissue close to the tracheal opening. A variant of this procedure is a tonsillectomy (often with adenoid removal) in children. In some adults, these growths at the back of the oral cavity also cause apnea and have been removed to good effect. Tissues of the pharynx near the tracheal opening have also been shrunk or removed by laser or cauterizing devices.

None of the foregoing noninvasive or invasive procedures is trivial, and each must be handled by qualified medical or dental personnel. Two behavioral approaches have been used and shown to be effective. Weight loss is one. Overweight patients with obstructive sleep apnea were able to reduce snoring and apneic episodes by losing weight, since the weight loss was accompanied by a reduction of excess fleshy tissue of the pharynx. Another simple procedure is to avoid sleeping on the back. In the supine position, fleshy tissue will tend to fall back and block the tracheal opening. Sleeping on one's side or "belly down" (prone) avoids this risk and reduces snoring and apnea. It has been found that alcohol, tobacco, and some sedatives, taken close to bedtime, make the opening to the airway susceptible to collapse, and thus avoiding these substances late in the day can have some efficacy.

Sudden Infant Death Syndrome (SIDS)

SIDS, also called crib death in the United States and cot death in other countries, claims the lives of more than 3000 apparently healthy babies each year in the United States. The incidence of SIDS is rare during the first month of life, peaks at 2 to 4 months of age, and then declines. A clear-cut cause of these deaths has not been established. There is evidence that sleeping in the prone position, especially with something soft in the crib, increases the chances of accidental asphyxiation due to the face being buried in the soft material with the nose and mouth being cut off from an outside source of air. Under these conditions, the infant is rebreathing an atmosphere of exhaled air that contains between 4 and 5% CO_2. Normal room air has a CO_2 concentration of 0.03 to 0.04% (Kemp and Thach, 1993; Skadberg and Markestad, 1997). Skadberg and Markestad (1997), in Bergen, Norway, who investigated conditions that could lead to this outcome, had previously reported that SIDS victims were usually found in the prone position with their heads covered by bedding material. They experimentally pursued that lead. They tested 30 infants at 2.5 months of age and 26 at 5 months of age. The younger infants were at an age that is most at risk for SIDS; the older ones were past that most vulnerable period. Under carefully monitored conditions, the heads of the sleeping infants were gently covered with a light quilt. In the supine position (face up), 23% of the 2.5 month-old infants were able to remove the cover from their faces and 60% did so at 5 months of age. In the prone position (belly down), none of the younger infants could manage to do that and only one 5 month old could. Clearly, sleeping face up without soft bedding, especially during the vulnerable period, is indicated.

In 1992 the American Academy of Pediatrics instituted a campaign called "Back to Sleep" in which they recommended that infants be put to sleep on their backs with pillows, quilts, and other soft materials removed from the cribs. The incidence of SIDS in the United States dropped from a high of over 5000 deaths per year to fewer than 3000.

However, SIDS is still the major cause of death for infants during the first year of life.

Another factor suggested as a cause of SIDS is related to the neural mechanism that controls the rate and depth of respiration. During periods of sleep when the rate and depth of respiration normally decrease, the blood levels of CO_2 increase and that of O_2 decrease. As described in the section on sleep apnea, these changes in blood gas concentration are sensed by receptors that signal the respiratory control region of the medulla to increase its activity. The resulting increased rate and depth of respiration rapidly correct the levels of CO_2 and O_2. It is well known that the nervous system in utero and in infancy is continually maturing, and there is recent evidence that in some infants the CO_2-sensitive neural mechanism in the lower brain stem that accomplishes the respiratory adjustments has not sufficiently matured to make the vital adjustments, with fatal results (Kinney et al., 2001; Bradley et al., 2002).

A combination of factors may contribute to SIDS. Infants who experience some degree of central or obstructive sleep apnea and also have difficulty in awakening may become SIDS victims. Ordinarily, an apneic episode during sleep is terminated by an arousal during which normal breathing is restored. Detailed discussions of SIDS and multiple possible causes are presented by Glotzbach et al. (1995) and Cornwell (1995).

A bizarre and tragic twist in the literature on SIDS is that some infant deaths attributed to SIDS are instead due to infanticide: the mother or other caregiver has suffocated the infant. A dramatic case is based on an article in the journal *Pediatrics* in which five children in a single family were reported to have died of sleep apnea (Steinschneider, 1972). Initially this story was presented as evidence for SIDS being a familial disease—that there is a heritable factor for SIDS. As described in the book, *The Death of Innocents* by Firstman and Talan (1997), it was later found that the mother had suffocated her children.

Narcolepsy

Narcolepsy is a neurological disorder with an incidence of occurrence in the United States of about 1 in 2000. It is less prevalent than Parkinson's disease and multiple sclerosis and about 10 times more frequent than amyotrophic lateral sclerosis (ALS, Lou Gehrig's disease) (J. M. Siegel, 2000a). The symptoms of narcolepsy certainly have been known for centuries, but a Parisian physician, Jean Baptiste Edouard Gelineau, first identified them as components of a neurological disorder in 1880. Gelineau's term, "narcolepsy," from the Greek, means sleep (narco) seizure (lepsy). Narcolepsy is sometimes mistaken for a form of epilepsy because the name carries the meaning of seizure and because in some narcoleptic attacks the person suddenly collapses to the floor.

The onset of the disorder typically occurs during the teenage or young adult years. The predominant symptom of narcolepsy, one may say the hallmark symptom, is excessive daytime sleepiness that culminates in a sleep attack. Narcolepsy is second only to sleep apnea as a cause of excessive daytime sleepiness. Narcoleptics experience frequent interruptions of their nighttime sleep (sleep fragmentation) that contribute to their daytime sleepiness. Three other symptoms, in addition to daytime sleepiness and sleep attacks, occur in narcolepsy. These are sudden episodes of paralysis during waking, called cataplexy*; paralysis during the transitions between sleep and waking, called sleep paralysis; and dreamlike hallucinations that occur during waking, called hypnogogic hallucinations. These three symptoms associated with narcolepsy resemble aspects of REM sleep—leading some workers to conclude that narcolepsy involves a dysfunction of REM sleep, as though REM sleep is breaking out into the waking state. This idea is related to the observation that narcoleptics have a very short latency for their first nighttime REM sleep episode. Whereas this normally occurs an hour to an hour and a half after falling asleep, narcoleptics have their first REM sleep episode within minutes of falling asleep. In addition to the major symptom, sleep attacks, narcoleptics usually show one or more of the other three symptoms, most frequently that of cataplexy. Only 20 to 25% of all narcoleptics develop all four symptoms.

Sleep attacks may occur frequently and at any time during the day. An attack often starts with the person becoming sleepy and progressing to the point where sleep is unavoidable. This usually occurs during monotonous, routine, or boring tasks. Attacks can also occur at patently inappropriate times such as during meals or while interacting with others. A sleep attack usually lasts a few minutes or in some cases as long as a half-hour. The person typically awakens fully rested, but after an hour or more another episode may recur.

About 75% of narcoleptics also exhibit cataplexy, sudden episodes of paralysis. Loss of muscle tonus to the point of paralysis is a normal component of REM sleep. But cataplexy is such an episode that in-

---

*Cataplexy should not be confused with catalepsy, a symptom of catatonic schizophrenia in which the limbs and trunk can be molded, like wax, into almost any position. There is no loss of muscle tonus in catalepsy.

trudes itself into the waking state. These episodes are as dramatic as the narcoleptic sleep attacks, if not more so. They often occur and appear to be triggered by a sudden heightened emotion, such as laughing at a joke's punch line or being surprised.

In cataplexy, the postural and antigravity muscles lose tonus. Sometimes this is rather mild and the facial and leg muscles relax so that the eyelids droop, the jaw sags, and the knees go a bit weak and bend. In other cases, the loss of tonus is complete (except for the muscles of respiration and of eye movements—as in REM sleep) and the person suddenly collapses to the floor in a heap and remains there completely limp for a few seconds to a number of minutes. The person, who is completely conscious during this time, may sleep—usually REM sleep—if he or she has fallen into a comfortable position.

The other two symptoms of narcolepsy are sleep paralysis and hypnogogic hallucinations. Sleep paralysis refers to the loss of muscle tonus, as during cataplexy and REM sleep, that occurs during the transition from waking to sleeping and from sleeping to waking. Nonnarcoleptics occasionally experience such episodes in sleep–waking transitions and sometimes cannot tell if they are awake or dreaming. If they are more awake than not, they may be frightened by the awareness that they cannot move. Narcoleptics who experience sleep paralysis more often are used to the phenomenon and wait it out more calmly.

Hypnogogic hallucinations are dreamlike experiences that occur when the person is falling asleep but is still awake and drowsy, and just after awakening. The hallucinations are sometimes vivid and frightening and are often accompanied by sleep paralysis.

The cause of narcolepsy and its associated symptoms has been a mystery for a long time, A major breakthrough occurred in the 1970s when it was discovered that dogs—and importantly, certain breeds of dogs—exhibited narcoleptic symptoms (Knecht et al., 1973; Mitler et al., 1974). As in many diseases, the development of an animal model marked the beginning of advances in discovering the underlying mechanisms of the disease. In 1977 Dement at Stanford University succeeded in breeding a colony of Doberman pinscher dogs that were particularly susceptible to narcoleptic attacks. Dement describes this project in a chapter on narcolepsy in his engagingly written book, *The Promise of Sleep* (Dement and Vaughan, 1999).

Narcoleptic dogs, like humans, exhibit more daytime sleep than normal dogs (suggesting excessive daytime sleepiness), fragmented nighttime sleep (frequent awakenings), and short latencies to REM sleep episodes. Most strikingly, narcoleptic dogs also display cataplexy. This can be triggered by the excitement of offering the dog a particularly desirable food, vigorously playing with the dog, or startling the dog. These episodes usually involve muscle weakness and sagging of the limbs and facial muscles for a few seconds. However, an episode of decreased muscle tonus may escalate into a full-blown cataplectic attack in which the animal suddenly falls to the floor in a heap. The dog appears to be conscious and is capable of following objects with eye movements. As in REM sleep, the extraocular muscles that control eye movements are not paralyzed.

An animal model in which animals are bred for the disease has led to two important directions of research. The fact that animals can be bred for narcolepsy indicates that a heritable factor is involved in its etiology, and has led to considerable research on the role of heritability in narcolepsy. Certainly the connection exists in canine narcolepsy, but heritability is less clear in human narcolepsy. The other line of research capitalizes on the fact that with an animal model, the neural mechanisms basic to this neurological disease can be investigated by means of invasive studies.

Recent research using animals has yielded information on the neural mechanisms of narcolepsy. In 1997 two presentations at the annual meeting of the Society for Neuroscience reported the presence of nerve cells in the hypothalamus that synthesize two related peptides later shown to be involved in narcolepsy. Within a year these peptide pairs had received two names. One name was orexin (orexins A and B for the related pair of peptides) and the other was hypocretin (hypocretin-1 and hypocretin-2). After some initial confusion, it was clearly established that orexin-A and hypocretin-1 are identical peptides and that orexin-B and hypocretin-2 are the same (Sakurai et al., 1998).*

In short order a number of breakthroughs in our understanding of narcolepsy occurred. The precise location of hypocretin cells in the hypothalamus was mapped; the anatomical projections of these cells to other brain regions were traced; the receptors on postsynaptic cells to

---

*It is of some interest to know how these exotic names were chosen. "Orexin" was used because early studies indicated that in rats, this peptide stimulated appetite, and the Greek root of the word literally means that. The term "hypocretin" is based on the findings that cells that synthesize the peptide are located in the hypothalamus and that the amino acid sequence is similar to that of the gut peptide secretin. Recent work has shown that nearly all the hypothalamic neurons that synthesize orexin (hypocretin) colocalize (also contain) another peptide transmitter, dynorphin. There is evidence that the narcolepsy influence is due to orexin and that feeding is regulated by dynorphin (Chou et al., 2001).

which these peptide transmitters bind were determined; the genes and associated chromosomes that predispose humans for narcolepsy and the gene that causes narcolepsy in dogs, were identified; and finally, a possible new treatment for narcolepsy was suggested.

In 1999 Masashi Yanagisawa at the University of Texas Medical Center in Dallas, working with a group of collaborators, used techniques of molecular genetics to disrupt the gene that codes for the amino acid sequence of the hypocretins in mice (Chemelli et al., 1999). This genetic manipulation produced "orexin knockout mice." The orexin/ hypocretin gene was, in a sense, deleted ("knocked out"). The knockout mice showed a number of the symptoms seen in narcoleptic humans and dogs, namely, periods of behavioral arrest that resembled cataplexy and increased bouts of sleep during usual waking periods.

In the same journal issue that described the work of Yanagisawa's group, Emmanuel Mignot at the Center for Narcolepsy at Stanford University, also working with collaborators, published a paper showing how the gene for canine narcolepsy is expressed to produce the narcoleptic symptoms (Lin et al., 1999). The gene for canine narcolepsy, aptly called *canarc-1*, located on chromosome 12, is described as having "high penetrance." This means that the gene has a high probability of being expressed (animals carrying the gene will, with a high probability, be narcoleptic). Mignot and his colleagues reported that canine narcolepsy was due to the deletion of certain amino acids of the protein that comprises the hypocretin-2 receptor. The receptor, in this truncated form, is rendered nonfunctional. The supposition is that the gene *canarc-1* causes the mutation that results in the nonfunctional hypocretin receptor. There is evidence that narcoleptic dogs bred for the trait have intact hypocretin cells in the hypothalamus but lack the functional receptors that permit the hypocretin transmitters to have their normal effects.

The case for human narcolepsy is more complicated. Here the evidence indicates that the cause is the loss of the hypothalamic cells that synthesize the hypocretins. A first indication of this was the finding that hypocretin could not be detected in the cerebrospinal fluid (CSF) of seven of nine narcoleptic patients and could be detected in the CSF of eight normal people (Nishino et al., 2000). This lead was followed up by the groups of J. M. Siegel at UCLA and Mignot at Stanford. Both groups used preserved brains of narcoleptic patients collected by Michael Aldrich, a neurologist at the University of Michigan Medical Center in Ann Arbor. Aldrich, the founder of the University of Michigan Sleep Disorder Laboratory, which had a major interest in the study and treatment of narcolepsy, died in July 2000 at the age of 51. Over the years he had accumulated a brain bank of preserved narcoleptic brains that could be used by scientists for detailed histological analysis. Before his tragic death, Aldrich had sent brain tissue from the narcoleptic brain bank to both the Siegel and Mignot groups for histological analysis. Both groups published their findings within weeks of each other (Peyron et al., 2000; Thannickal et al., 2000). Each paper was dedicated to the memory of Michael Aldrich.

The findings of the two groups, for the most part, were congruent. The brains of human narcoleptics showed a profound reduction in hypocretin. Siegel's group reported a reduction of 85 to 90% in the number of hypocretin neurons. Neurons of other types located in the same area of the hypothalamus and interspersed among the hypocretin cells were not affected. This showed that whatever caused the loss of hypocretin cells was specific for those cells and did not destroy cells of other types in the same region. The two reports were discrepant in that the Mignot group did not find signs of gliosis in the hypothalamus, whereas the Siegel group did. "Gliosis" refers to the proliferation of astrocytes, a type of glial cell, which occurs as a reaction to nerve cell damage and nerve death. Siegel and coworkers reported gliosis in the region that showed hypocretin cell loss. Loss of cells and gliosis may be due to a number of things, including an immune or autoimmune reaction, a toxin, or a bacterial or viral infection.

It is now important to determine what causes the destruction of hypocretin cells in human narcoleptics. In dogs we know that a heritable genetic mutation renders hypocretin receptors nonfunctional and is the cause of canine narcolepsy. In humans the damage is to the hypocretin cells themselves. Is there a genetic factor operating here too? The answer is not simple. There is evidence for a heritable influence, but not as strong an influence as that demonstrated in canine narcolepsy. The disease in humans has been found to run in families, but this is not common. An indication of the degree of heritability comes from identical twin studies, If one twin is narcoleptic, there is only a 25 to 30% chance that the other twin will also be narcoleptic (Mignot, 1998). Since the incidence of the disease is 1 in 2000, finding two narcoleptics in the same family is highly unlikely unless there is a genetic factor involved—or an environmental factor highly specific to that family. So even though twins share 100% of their genes, the chances are that only one of the twins will suffer narcolepsy. An environmental triggering influence may cause a genetic factor to be expressed in one case and not in the other. A comment by a speaker at a toxicology conference is apt here: "Genetics loads the gun; environment pulls the trigger."

The search for one or more genes linked to human narcolepsy has shown that the family of genes called human leukocyte antigens (HLAs), located on chromosome 6, code for mechanisms that trigger an immune response. In the case of narcolepsy, it is thought that particular variants of the HLA gene trigger an autoimmune response. That is, the body's immune system attacks its own tissue as though it is a foreign protein (like disease-causing bacteria). The gene most closely associated with narcolepsy has been identified as HLA DBQ1*0602 (Mignot, 2000). A number of autoimmune diseases are known to be linked to specific HLA variants; examples are multiple sclerosis, myasthenia gravis, and Graves' disease (hyperthyroidism). In all cases the force of the immune system is directed quite specifically only to certain tissue. In human narcolepsy, the supposition is that the autoimmune attack is upon the hypocretin cells of the hypothalamus.

Evidence for this hypothesis is that more than 90% of narcoleptics with cataplexy test positive for the presence of the HLA gene DBQ*0602. However, about 23% of nonnarcoleptics also demonstrate this gene, and a few narcoleptic patients do not test positive for the HLA marker (Mignot, 2000). It has been suggested that the expression of the gene for an autoimmune attack may be influenced by certain environmental events (e.g., a virus or toxin) (Thannickal et al., 2000). The demonstration by Siegel's group of the presence of gliosis in the region of the narcoleptic brains that normally has hypocretin cells is consistent with a process that destroyed those cells—possibly an autoimmune process. Additional support from Siegel's group for an autoimmune response was presented recently at a neuroscience conference: narcoleptic dogs treated with immunosuppressant drugs doubled the age of cataplexy onset and decreased the time spent in cataplexy by more than 85% compared with controls (Boehmer and Siegel, 2001).

Recent work offers some promise for a treatment for narcolepsy. Since human narcoleptics have a loss of hypocretin-producing cells and therefore the transmitter itself, perhaps the administration of hypocretin would reverse the narcoleptic symptoms. For human clinical use, it is not feasible to inject a substance directly into the brain; that is, the hypocretin would have to be administered systemically. For hypocretin to have a central nervous system effect, it would have to cross the blood–brain barrier and bind to hypocretin receptors in the brain. However, shortly after hypocretin was discovered, studies reported that systemically administered hypocretin did not cross the blood–brain barrier in sufficient quantity to be effective. And as noted, injecting the transmitter directly into the brain or a ventricle is not a viable clinical option. Using information provided by Kastin and Akerstrom (1999), however, workers in Siegel's laboratory at UCLA succeeded in systemically administering hypocretin-1 so that it passed the blood–brain barrier. Narcoleptic Doberman pinscher dogs were used as subjects (John et al., 2000).

The findings were encouraging. Hypocretin reduced and in some cases abolished cataplectic attacks in a dose-dependent manner. Three of the dogs in the experiment were given repeated doses over a period of days. Two of the dogs after 3 days of dosing showed a complete absence of cataplexy for 3 successive days. A third dog with more severe cataplexy required 5 days of treatment to produce a suppression of cataplectic attacks. The authors point out that these three dogs had been observed for 35 consecutive days prior to treatment, and an absence of cataplexy had never been observed on any of those days. The severity of cataplectic attacks gradually returned after treatment days ended.

Another effect of hypocretin treatment was the consolidation of sleep and waking periods. As described earlier, narcoleptic humans and dogs suffer from multiple intrusions of sleep episodes during the day and multiple awakenings during the night. During treatment days and succeeding days, the dogs had fewer sleep bouts during the day and fewer awakenings during the night; that is, their sleep and waking periods were no longer as fragmented as they had been during pretreatment days.

The interpretation of these findings is complicated by data that canine narcolepsy is due to a genetic mutation that interferes with the normal synthesis of the hypocretin-2 receptor. So how does the administration of hypocretin-1 work to have its therapeutic effect? John et al. (2000) suggest that the hypocretin-2 receptor that is altered in narcoleptic dogs may still be responsive, but at a reduced level. Also proposed is the possibility that activation of hypocretin-1 receptors, or other receptors not yet identified, may be the cause of the therapeutic effect of hypocretin-1 administration.

The application of these findings to the treatment of human narcolepsy is complicated by observations that canine narcolepsy is caused by the disruption of hypocretin receptors and human narcolepsy is caused by the loss of the hypocretin neurons themselves. An animal model of narcolepsy based on the loss of hypocretin cells (e.g., hypocretin knockout animals) would be closer to the human narcoleptic condition. Such an animal model that tests for the efficacy of hypocretin replacement treatment would be more directly applicable to human use.

References

Adametz JH (1959) Rate of recovery of functioning in cats with rostral reticular lesions: an experimental study. J Neurosurg 16:85–97.

Adey WR, Segundo JP, Livingston RB (1957) Corticofugal influences on intrinsic brain stem conduction in cat and monkey. J Neurophysiol 20:1–16.

Adey WR, Kado RT, Rhodes JM (1963) Sleep, cortcal and subcortical recordings in the chimpanzee. Electroencephalogr Clin Neurophysiol 28:368–373.

Albert IB, Cicala GA, Siegel J (1970) The behavioral effects of REM sleep deprivation in rats. Psychophysiology 6:550–560.

Allison T, van Twyer H, Goff WR (1972) Electrophysiological studies of the echidna, *Tachyglossus aculeatus*. I. Waking and sleeping. Arch Ital Biol 110:145–184.

Allison T, Gerber SD, Breedlove SM, Dryden GL (1977) A behavioural and polygraphic study of sleep in the shrews *Suncus murinus, Blarina brevicauda* and *Cryptatis parva*. Behav Biol 20:354–366.

Aserinsky E (1996) The discovery of REM sleep. J Hist Neurosci 5:213–227.

Aserinsky E, Kleitman N (1953) Regularly occurring periods of eye motility, and concomitant phenomena, during sleep. Science 118:273–274.

Batini C, Moruzzi G, Palestini M, Rossi GF, Zanchetti A (1958) Persistent patterns of wakefulness in the pretrigeminal midpontine preparation. Science 128:30–32.

Batini C, Magni CF, Palestini M, Rossi GF, Zanchetti A (1959) Neural mechanisms underlying the enduring EEG and behavioral activation in the midpontine pretrigeminal cat. Arch Ital Biol 97:13–25.

Beck A (1891) Determination of localization in the brain and spinal cord by means of electrical phenomena. Pol Akad Umiejetni, Ser II, 1:186–232.

Berger H (1929) Über das Elektrenkephalogramm. Arch Psychiatr Nervenkrank 87:527–570.

Bergmann BM, Kushida CA, Everson CA, Gilliland MA, Overmyer WH, Rechtschaffen A (1989) Sleep deprivation in the rat. II. Methodology. Sleep 12:5–12.

Bergmann BM, Gilliland MA, Feng PF, Russell DA, Shaw P, Wright M, Rechtschaffen A, Alverdy JC (1996) Are physiological effects of sleep deprivation in the rat mediated by bacterial invasion? Sleep 19:554–562.

Boehmer LN, Siegel JM (2001) Treatment affecting the course of canine narcolepsy. Soc Neurosc Abstr 27: program number 8.4.

Borbély AA (1982) A two process model of sleep regulation. Hum Neurobiol 1:195–204.

Borbély AA, Achermann P (2000) Sleep homeostasis and models of sleep regulation. In: Principles and practice of sleep medicine, 3rd ed (Kryger MH, Roth T, Dement WC, eds), pp 377–390. Philadelphia: WB Saunders.

Bradley PB, Mollica A (1958) The effect of adrenaline and acetylcholine on single unit activity in the reticular formation of the decerebrate cat. Arch Ital Biol 96:168–186.

Bradley SR, Pieribone VA, Wang W, Severson A, Jacobs RA, Richerson GB (2002) Chemosensitive secotonergic neurons are closely associated with large medullary arteries. Nature Neurosci 5:401–402.

Braun AR, Balkin TJ, Wesensten NJ, Carson RE, Varga M, Baldwin P, Selbie S, Belenky G, Herscovitch P (1997) Regional cerebral blood flow throughout the sleep–wake cycle. Brain 120:1173–1197.

Brazier MAB (1961) A history of the electrical activity of the brain. London: Pitman Medical Publishing.

Brazier MAB (1980) Trails leading to the concept of the ascending reticular system: the state of knowledge before 1949. In: The reticular formation revisited: specifying function for a nonspecific system (Hobson JA, Brazier MAB, eds), pp 331–352. New York: Raven Press.

Bremer F (1935) Cerveau "isolé" et physiologie du sommeil. C R Soc Biol 118: 1235–1241.

Bremer F (1936) Nouvelles recherches sur le mécanisme du sommeil. C R Soc Biol 122:460–464.

Bremer F (1938) L'activité électrique de l'écorce cérébrale et le problème physiologique du sommeil. Bull Soc Ital Biol Sper 13:271–290.

Bremer F (1970) Preoptic hypnogenic focus and mesencephalic reticular formation. Brain Res 21:132–134.

Bremer F (1974) Historical development of ideas on sleep. In: Basic sleep mechanisms (Petre-Quadens O, Schlag J, eds), pp 3–11. New York: Academic Press.

Buchwald NA, Heuser G, Wyers EJ, Lauprecht CW (1961a) The "caudate spindle." III. Inhibition by high frequency stimulation of subcortical structures. Electroencephalogr Clin Neurophysiol 13:525–530.

Buchwald NA, Wyers EJ, Lauprecht CW, Heuser G (1961b) The "caudate spindle." IV. A behavioral index of caudate-induced inhibition. Electroencephalogr Clin Neurophysiol 13:531–537.

Buchwald NA, Wyers EJ, Okuma T, Heuser G (1961c) The "caudate spindle." I. Electrophysiological properties. Electroencephalogr Clin Neurophysiol 13:509–518.

Campbell S, Tobler I (1984) Animal sleep: a review of sleep duration across phylogeny. Neurosci Biobehav Rev 8:269–300.

Caton R (1875) The electrical currents of the brain. Br Med J 2:278.

Cauter EV, Spiegel K (1999) Circadian and sleep control of hormonal secretions. In: Regulation of sleep and circadian rhythms (Turek FW, Zee PC, eds), pp 397–425. New York: Marcel Dekker.

Chase MH, Morales FR (2000) Control of motoneurons during sleep. In: Principles and practice of sleep medicine, 3rd ed (Kryger MH, Roth T, Dement WC, eds), pp 155–168. Philadelphia: WB Saunders.

Chemelli RM, Willie JT, Sinton CM, Elmquist JK, Scammell T, Lee JK, Richardson JA, Williams SC, Xiong Y, Kisanuhki RE, Fitch TE, Nakazato M, Hammer RE, Saper

CB, Yanagisawa M (1999) Narcolepsy in *orexin* knockout mice: molecular genetics of sleep regulation. Cell 98:437–451.

Chou TC, Lee CE, Lu J, Elmquist JK, Hara J, Willie JT, Beuckmann CT, Chemelli RM, Sakurai T, Yanagisawa M, Saper CB, Scammell TE (2001) Orexin (hypocretin) neurons contain dynorphin. J Neurosci 21:RC168:1–6.

Chrobak JJ, Buzsáki G (1998) Gamma oscillations in entorhinal cortex of freely behaving rat. J Neurosci 18:388–398.

Cicala GA, Albert IB, Ulmer FA (1970) Sleep and other behaviours of the red kangaroo (*Megaleia rufa*). Animal Behav 18:786–790.

Clarke RH, Henderson EE (1912) Atlas of photographs of sections of the frozen cranium and brain of the cat (*Felis domestica*). J Psychol Neurol 18:391–409.

Clarke RH, Horsley V (1906) A method of investigating the deep ganglia and tracts of the central nervous system (cerebellum). Br Med J 2:1799–1800.

Clemente CD (1968) Forebrain mechanisms related to internal inhibition and sleep. Conditional Reflex 3:145–174.

Cohen HB, Dement WC (1965) Sleep: changes in threshold to electroconvulsive shock in rats after deprivation of "paradoxical" phase. Science 150:1318–1319.

Cohen HB, Duncan R, Dement WC (1967) The effect of electroconvulsive shock in cats deprived of REM sleep. Science 156:1646–1648.

Cohen HB, Thomas J, Dement WC (1970) Sleep stages, REM deprivation and electroconvulsive threshold in the cat. Brain Res 19:313–317.

Corner MA, Bour HL (1984) Postnatal development of spontaneous neuronal discharge in the pontine reticular formation of free-moving rats during sleep and wakefulness. Exp Brain Res 54:66–72.

Cornwell AC (1995) Sleep and sudden infant death syndrome. In: Clinical handbook of sleep disorders in children (Schaefer CE, ed), pp 15–47. Northvale, NJ: Jason Aronson.

Dahlstrom A, Fuxe K (1964) Evidence for the existence of monoamine-containing neurons in the central nervous system. I. Demonstration of monoamines in the cell bodies of brain stem neurons. Acta Physiol Scand, suppl 232:1–55.

Datta S (1995) Neuronal activity in the peribrachial area: relationship to behavioral state control. Neurosci Biobehav Rev 19:67–84.

Datta S, Hobson JA (1994) Neuronal activity in the caudo-lateral peribrachial pons: relationship to PGO waves and rapid eye movements, J Neurophysiol 71:95–109.

Datta S, Pare D, Oakson G, Steriade M (1989) Thalamic-projecting neurons in brainstem cholinergic nuclei increase their firing rates one minute in advance of EEG desynchronization associated with REM sleep. Soc Neurosci Abstr 15:452.

Davenne D, Adrien J (1984) Suppression of PGO waves in the kitten: anatomical effects on the lateral geniculate nucleus. Neurosci Lett 45:33–38.

Dell P, Bonvallet M, Hugelin A (1961) Mechanisms of reticular deactivation. In: A Ciba Foundation symposium on the nature of sleep (Wolstenholme GFW, O'Conner MC, eds), pp 86–102. London: Churchill.

Dement W (1955a) Dream recall and eye movements during sleep in schizophrenics and normals. J Nerv Ment Dis 122:263–269.

Dement W (1955b) Rapid eye movements during sleep in schizophrenics and nonschizophrenics and their relation to dream recall. Doctoral dissertation, University of Chicago.

Dement W (1958) The occurrence of low voltage, fast electroencephalogram patterns during behavioral sleep in the cat. Electroencephalogr Clin Neurophysiol 10:291–296.

Dement WC (1960) The effect of dream deprivation. Science 131:1705–1707.

Dement WC (1965) An essay on dreams: the role of physiology in understanding their nature. In: New directions in psychology, II (Newcomb TM, ed), pp 135–257. New York: Holt, Rinehart & Winston.

Dement WC (1976) Some must watch while some must sleep. New York: WW Norton.

Dement WC, Fisher C (1963) Experimental interference with the sleep cycle. Can Psychiatr Assoc J 8:400–405.

Dement WC, Kleitman N (1957a) The relation of eye movements during sleep to dream activity: an objective method for the study of dreaming. J Exp Psychol 53:339–346.

Dement WC, Kleitman N (1957b) Cyclic variations in EEG during sleep and their relation to eye movements, body motility, and dreaming. Electroencephalogr Clin Neurophysiol 9:673–690.

Dement WC, Vaughan C (1999) The promise of sleep. New York: Delacorte Press.

Dempsey EW, Morison RS (1942a) The production of rhythmically recurrent cortical potentials after localized thalamic stimulation. Am J Physiol 135:293–300.

Dempsey EW, Morison RS (1942b) The interaction of certain spontaneous and induced cortical potentials. Am J Physiol 135:301–308.

Dempsey EW, Morison RS (1943) The electrical activity of the thalamocortical relay system. Am J Physiol 138:283–296.

Eiland MM, Ramanathan L, Gulyani S, Gilliland M, Bergmann BM, Rechtschaffen A, Siegel JM (1999) Sleep deprivation related changes in the anterior hypothalamus of sleep deprived rats. Sleep Res Online 2 (suppl 1):527.

El-Kafi B, Leger L, Seguin S, Jouvet M, Cespuglio R (1995) Sleep permissive components within the dorsal raphé nucleus in the rat. Brain Res 686:150–159.

Eriksson KS, Sergeeva O, Brown RE, Haas HL (2001) Orexin excites the histaminergic tuberomammillary neurons. Soc Neurosci Abstr 27: program number 8.7.

Erlanger J, Gasser HS (1924) The compound nature of the action current of nerve as disclosed by the cathode ray oscillograph. Am J Physiol 70:624–666.

Erlanger J, Gasser HS (1937) Electrical signs of nervous activity. Philadelphia: University of Pennsylvania Press.

Everson CA (1993) Sustained sleep deprivation impairs host defense. Am J Physiol Regul Integ Comp Physiol 265:R1148-R1154.

Everson CA, Toth LA (2000) Systemic bacterial invasion induced by sleep deprivation. Am J Physiol Regul Integ Comp Physiol 278:R905-R916.

Faure J, Vincent D, LeNovenne J, Geissmann P (1963) Sommeil lent et stade paradoxal chez le lapin des deux sexes: role du milieu. C R Hebd Seances Soc Biol 157:799–804.

Firstman R, Talan J (1997) The death of innocents. New York: Bantam Books.

Fishbein W, Gutwein BM (1977) Paradoxical sleep and memory storage processes. Behav Biol 19:425–464.

Fishbein W, Gutwein BM (1981) Paradoxical sleep and a theory of long-term memory. In: Sleep, dreams and memory (Fishbein W, ed), pp 147–182. Jamaica, NY: Spectrum Publications.

Fisher C, Dement WC (1963) Studies in the psychopathology of sleep and dreams. Am J Psychiatr 119:1160–1168.

Foster M (1901) Lectures on the history of physiology during the sixteenth, seventeenth and eighteenth centuries. Cambridge: Cambridge University Press.

Frank MG, Issa NP, Stryker MP (2001) Sleep enhances plasticity in the developing visual cortex. Neuron 30:275–287.

Friedmann J, Globus G, Huntley A, Mullaney D, Naitoh P, Johnson L (1977) Performance and mood during and after gradual sleep reduction. Psychophysiology 14:245–250.

Gais S, Plihal W, Wagner U, Born J (2000) Early sleep triggers memory for early discrimination skills. Neurosci 3:1335–1339.

Gasser HS, Erlanger J (1922) A study of the action currents of nerve with the cathode ray oscillograph. Am J Physiol 62:496–524.

Glotzbach SF, Ariagno RL, Harper RM (1995) Sleep and the sudden infant death syndrome. In: Principles and practice of sleep medicine in the child (Ferber R, Kryger M, eds), pp 231–244. Philadelphia: WB Saunders.

Hartmann E, Bernstein J, Wilson C (1967) Sleep and dreaming in the elephant. Psychophysiology 4:389.

Hendricks JC, Morrison AR, Mann GL (1982) Different behaviors during paradoxical sleep without atonia depend on pontine lesion site. Brain Res 239:85–105.

Hendricks JC, Stefanie MF, Panckeri KA, Chavkin J, Williams JA, Sehgal A, Pack A (2000) Rest in *Drosophila* is a sleep-like state. Neuron 25:129–138.

Hennevin E, Hars B, Maho C, Bloch V (1995) Processing of learned information in paradoxical sleep: relevance for memory. Behav Brain Res 69:125–135.

Hernandez-Peon R, Chavez-Ibarra G, Morgane PJ, Timo-Iaria C (1963) Limbic cholinergic pathways involved in sleep and emotional behavior. Exp Neurol 8:93–111.

Hess WR (1944) Das Schlafsyndrom als Folge diencephaler Reizung. Helv Physiol Pharmacol Acta 2:305–344.

Hess WR (1954) The diencephalic sleep centre. In: Brain mechanisms and consciousness (Bremer F, Japer HH, eds), pp 117–136. Oxford: Blackwell.

Hess WR (1957) The functional organization of the diencephalon. New York: Grune & Stratton.

Heuser G, Buchwald NA, Wyers EJ (1961) The "caudate spindle." II. Facilitatory and inhibitory caudate–cortical pathways. Electroencephalogr Clin Neurophysiol 13: 519–524.

Hobson JA (1988) The dreaming brain. New York: Basic Books.

Hobson JA, McCarley RW, Pivik T, Freedman R (1974) Selective firing by cat pontine brain stem neurons in desynchronized sleep. J Neurophysiol 37:497–511.

Hobson JA, McCarley RW, Wyzinski PW (1975) Sleep cycle oscillation: reciprocal discharge by two brain stem neuronal groups. Science 189:55–58.

Hobson JA, Stickgold R, Pace-Schott EF (1998) The neuropsychology of REM sleep dreaming. NeuroReport 9:R1–R14.

Hopkins DA, Darvesh S, de Groot MHM, Rusak B (2001) Orexin immunoreactivity in normal and Alzheimer's disease brainstem. Soc Neurosci Abstr 27: program number 965.11.

Horne J (1988) Why we sleep: the function of sleep in humans and other animals. Oxford: Oxford University Press.

Hugelin A, Bonvallet M (1957a) Tonus cortical et contrôle de la facilitation motrice d'origine réticulaire. J Physiol (Paris) 49:1171–1200.

Hugelin A, Bonvallet M (1957b) Étude expérimentale des interrelations reticulocorticales. Proposition d'une théorie de l'asservissement réticulaire à un système diffus cortical. J Physiol (Paris) 49:1201–1223.

Hugelin A, Bonvallet M (1957c) Analyse des post-décharges réticulaires et corticales engendrées par des stimulations électriques réticulaires. J Physiol (Paris) 49:1225–1251.

Hugelin A, Bonvallet M (1958) Effets moteurs et corticaux d'origine réticulaire au cours des stimulations somesthésiques. Rôle des interactions cortico-réticulaires dans le déterminisme du réveil. J Physiol (Paris) 50:951–977.

Huttenlocker PR (1961) Evoked and spontaneous activity in single units of medial brain stem during natural sleep and waking. J Neurophysiol 24:451–468.

Inoué S (1989) Biology of sleep substances. Boca Raton, FL: CRC Press.

Inoué S, Honda K, Kimura M, Okano Y, Sun J, Ikeda M, Sagara M, Azuma S, Kodama T, Saha U (1997) A function of sleep: neuronal detoxification in the brain? In: Sleep and sleep disorders: from molecule to behavior (Hayaishi O, Inoué S, eds), pp 401–415. Tokyo: Academic Press.

Ishimori K (1909) True cause of sleep: a hypnogenic substance as evidenced in the brain of sleep-deprived animals. Tokyo Igakkai Zasshi 23:429–457.

Jacobs BL, Asher R, Dement WC (1973) Electrophysiological and behavioral effects of electrical stimulation of raphé nuclei in cats. Physiol Behav 11:489–495.

Jasper HH (1961) Thalamic reticular system. In: Electrical stimulation of the brain (Sheer DE, ed) pp 277–287. Austin: University of Texas Press.

John J, Wu M-F, Siegel JM (2000) System administration of hypocretin-1 reduces cataplexy and normalizes sleep and waking durations in narcoleptic dogs. Sleep Res Online 3:23–28.

Jones BE (1989) Basic mechanisms of sleep–wake states. In: Principles and practice of sleep medicine (Kryger MH, Roth T, Dement WC, eds), pp 121–138. Philadelphia: WB Saunders.

Jones BE (1990) Influence of the brainstem reticular formation, including intrinsic monoaminergic and cholinergic neurons, on forebrain mechanisms of sleep and waking. In: The diencephalon and sleep (Mancia M, Marini G, eds), pp 31–48. New York: Raven Press.

Jones BE (1993) The organization of cholinergic systems and their functional importance in sleep–waking states. In: Progress in brain research, vol 98, Cholinergic function and dysfunction (Cuello AC, ed), pp 61–71. Amsterdam: Elsevier.

Jones BE (2000) Basic mechanisms of sleep–wake states. In: Principles and practice of sleep medicine, 3rd ed (Kryger MH, Roth T, Dement WC, eds), pp 134–154. Philadelphia: WB Saunders.

Jones BE, Beaudet A (1987) Distribution of acetylcholine and catecholamine neurons in the cat brain stem studied by choline acetyltransferase and tyrosine hydroxylase immunohistochemistry. J Comp Neurol 261:15–32.

Jones BE, Webster HH (1988) Neurotoxic lesions of the dorsolateral pontomesencephalic tegmentum–cholinergic cell area in the cat. I. Effects upon the cholinergic innervation of the brain. Brain Res 451:13–32.

Jones BE, Yang TZ (1985) The efferent projections from the reticular formation and the locus coeruleus studied by anterograde and retrograde axonal transport in the rat. J Comp Neurol 242:56–92.

Jones JS (1995) Hans Berger. In: Notable twentieth century scientists, vol 1 (McMurray EJ, ed), pp 152–154. Detroit: Gale.

Jouvet M (1962) Recherches sur les structures nerveuses et le mécanismes respon-sables des différentes phases du sommeil physiologique. Arch Ital Biol 100:125–206.

Jouvet M (1972) The role of monoamine and acetylcholine-containing neurons in the regulation of the sleep–waking cycle. Ergebn Physiol 64:166–307.

Jouvet M (1996) The mechanisms of waking: from the mesencephalic reticular system to multiple networks. Arch Physiol Biochem 104:762–769.

Jouvet-Mounier D, Astic L, Lacote D (1970) Ontogenesis of the states of sleep in rat, cat, and guinea pig during the first postnatal month. Dev Psychobiol 2:216–239.

Jouvet-Mounier D, Vimont P, Delorme JF, Jouvet M (1964) Étude de la privation de phase paradoxal du sommeil chez le chat. C R Soc Biol (Paris) 158:756–759.

Jouvet M, Buda C, Sastre PC (1995) Is a pacemaker responsible for the ultradian rhythm of paradoxical sleep? Arch Ital Biol 134:39–56.

Jouvet M, Michel F, Courjon J (1959) Sur un stade d'activité électrique cérébrale rapide au cours du sommeil physiologique. C R Soc Biol 153:1024–1028.

Jouvet M, Valatx JL (1962) Étude polygraphique du sommeil chez l'agneau. C R Soc Biol (Paris) 156:1411–1414.

Kaada BR, Johannessen NB (1960) Generalized electrocortical activation by cortical stimulation in the cat. Electroencephalogr Clin Neurophysiol 12:567–573.

Karni A, Sagi D (1991) Where practice makes perfect in texture discrimination: evi-dence for primary visual cortex plasticity. Proc Natl Acad Sci USA 88:4966–4970.

Karni A, Tanne D, Rubenstein BS, Askenasy JJ, Sagi D (1994) Dependence on REM sleep of overnight improvement of a perceptual skill. Science 265:679–682.

Kastin AJ, Akerstrom V (1999) Orexin A but not orexin B rapidly enters brain from blood by simple diffusion. J Pharmacol Exp Ther 289:219–223.

Kemp JS, Thach BT (1993) A sleep position-dependent mechanism for infant death on sheepskins. Am J Dis Child 147:642–646.

Kilduff TS, Peyron C (2000) The hypocretin/orexin ligand–receptor system: implica-tion for sleep and sleep disorders. Trends Neurosci 23:359–365.

Kinney HC, Filiano JJ, White WF (2001) Medullary serotonergic network deficiency in the sudden infant death syndrome: Review of a 15-year study of a single data set. J Neuropath Exp Neurol 60:228–247.

Kleitman N (1963) Sleep and wakefulness. Chicago: University of Chicago Press.

Knecht C, Oliver JE, Redding R, Selcer R, Johnson G (1973) Narcolepsy in a dog and a cat. J Am Vet Med Assoc 162:1052–1053.

Kripke DF, Garfinkel L, Wingard DL, Klauber MR, Marler MR (2002) Mortality asso-ciated with sleep duration and insomnia. Arch Gen Psychiatr 59:131–136.

Krueger J, Walter J, Levin C (1985) Factor S and related somnogens: an immune the-ory for slow-wave sleep. In: Brain mechanisms of sleep (McGinty D, Drucker-Colín R, Morrison A, Parmeggiani L, eds), pp 253–275. New York: Raven Press.

Kryger MH, Roth T, Dement WC (eds) (2000) Principles and practice of sleep medi-cine, 3rd ed. Philadelphia: WB Saunders.

Kushida CA, Bergmann BM, Rechtschaffen A (1989) Sleep deprivation in the rat. IV. Paradoxical sleep deprivation. Sleep 12:22–30.

Latash LP, Galina GS (1975) Polygraphic characteristics of the dog's sleep. Sleep Res 4:145.

Lavie P (1996) The enchanted world of sleep. New Haven, CT: Yale University Press.

LeSauter J, Silver R (1999) Localization of a suprachiasmatic nucleus subregion regu-lating locomotor rhythmicity. J Neurosci 19:5574–5585.

Lin L, Faraco J, Li R, Kadotani H, Rogers W, Lin X, Qui X, de Jong PJ, Nishino S, Mignot E (1999) The sleep disorder canine narcolepsy is caused by a mutation in the *hypocretin (orexin) receptor 2* gene. Cell 98:365–376.

Lin JS, Hou Y, Sakai K, Jouvet M (1996) Histaminergic descending inputs to the mesopontine tegmentum and their role in the control of cortical activation and wakefulness in the cat. J Neurosci 16:1523–1537.

Lin JS, Sakai K, Jouvet M (1988) Evidence for histaminergic arousal mechanisms in the hypothalamus of cat. Neuropharmacology 27:111–122.

Lindsley DB, Bowden J, Magoun HW (1949) Effect upon the EEG of acute injury to the brain stem activating system. Electroencephalogr Clin Neurophysiol 1:475–486.

Lindsley, DB, Schreiner LH, Knowles WB, Magoun HW (1950) Behavior and EEG changes following chronic brain stem lesions in the cat. Electroencephalogr Clin Neurophysiol 2:483–498.

Lineberry CG, Siegel J (1971) EEG synchronization, behavioral inhibition, and mesencephalic unit effects produced by stimulation of orbital cortex, basal forebrain and caudate nucleus. Brain Res 34:143–161.

Louie K, Wilson MA (2001) Temporally structured replay of awake hippocampal ensemble activity during rapid eye movement sleep. Neuron 29:145–156.

Macchi G, Bentivoglio M (1986) The thalamic intralaminar nuclei and the cerebral cortex. In: Cerebral cortex, vol 5, Sensory-motor areas and aspects of cortical connectivity (Jones EG, Peters A, eds), pp 355–401. New York: Plenum Press.

Magnes J, Moruzzi G, Pompeiano O (1961a) Electroencephalogram-synchronizing structures in the lower brain stem. In: The nature of sleep (Wolstenholme GFW, O'Conner MC, eds), pp 57–78. London: Churchill.

Magnes J, Moruzzi G, Pompeiano O (1961b) Synchronization of EEG produced by low-frequency electrical stimulation of the region of the solitary tract. Arch Ital Biol 99:33–67.

Magoun HW (1954) The ascending reticular system and wakefulness. In: Brain mechanisms and consciousness (Delafresnaye JF, ed), pp 1–20. Springfield, IL: Charles C Thomas.

Magoun HW (1963) The waking brain. Springfield: Charles C Thomas.

Mancia M, Mariotti M, Spreafico R (1974) Caudo-rostral brainstem reciprocal influences in the cat. Brain Res 80:41–51.

Maquet P, Dive D, Salmon E, Sadzot B, Franco G, Poirrier R, von Frenkell R, Franck G (1990) Cerebral glucose utilization during sleep–wake cycle in man determined by positron emission tomography and [^{18}F]2-fluoro-2 deoxy-D-glucose method. Brain Res 513:136–143.

Maquet P, Peters J, Aerts J, Delfiore G, Degueldre C, Luxen A, Franck G (1996) Functional neuroanatomy of human rapid-eye movement sleep and dreaming. Nature 383:163–166.

Maquet P, Laureys S, Peigneux P, Fuchs S, Petiau C, Phillips C, Aerts J, Del Fiore G, Degueldre C, Meulemans T, Luxen A, Franck G, Van Der Linden M, Smith C, Cleeremans A (2000) Experience-dependent changes in cerebral activation during human REM sleep. Nat Neurosci 3:831–836.

Marshall LH, Magoun HW (1990) The Horsley–Clarke stereotaxic instrument: the beginning. Kopf Carrier, October 1990, 1–5.

Marshall LH, Magoun HW (1991) The Horsley–Clarke stereotaxic instrument: the first three insruments. Kopf Carrier, May 1991, 1–5.

Massopust LC, Jr (1961) Stereotaxic atlases: A. Diencephalon of the rat. In: Electrical stimulation of the brain (Sheer DE, ed), pp 182–202. Austin: University of Texas Press.

McCarley RW, Hobson JA (1975) Neuronal excitability modulation over the sleep cycle: a structural and mathematical model. Science 189:58–60.

McCormick DA, Bal T (1997) Sleep and arousal: thalamocortical mechanisms. Annu Rev Neurosci 20:185–215.

McGrath MJ, Cohen DB (1978) REM sleep facilitation of adaptive waking behavior: a review of the literature. Psychol Bull 85:24–57.

Mignot E (1998) Genetic and familial aspects of narcolepsy. Neurol 50 (suppl 1): 816–822.

Mignot E (2000) Pathophysiology of narcolepsy. In: Principles and practice of sleep medicine, 3rd ed (Kryger MH, Roth T, Dement WC, eds), pp 663–675. Philadelphia: WB Saunders.

Miltner WHR, Braun C, Arnold M, Witte H, Taub E (1999) Coherence of gamma-band EEG activity as a basis for associative learning. Nature 397:434–436.

Mirmiran M (1995) The function of fetal/neonatal rapid eye movement sleep. Behav Brain Res 69:13–22.

Mirmiran M, Van Someran E (1993) The importance of REM sleep for brain maturation. J Sleep Res 2:188–192.

Mitler MM, Boysen M, Campbell L, Dement WC (1974) Narcolepsy–cataplexy in a female dog. Exp Neurol 45:332–340.

Monnier M, Koller T, Graber S (1963) Humoral influences of induced sleep and arousal upon electrical brain activity of animals with crossed circulation. Exp Neurol 8:264–277.

Moore RY (1999) Circadian timing. In: Fundamental neuroscience (Zigmond MJ, Bloom FE, Landis SC, Roberts JL, Squire LR, eds), pp 1189–1206. San Diego, CA: Academic Press.

Moore RY, Eichler VB (1972) Loss of a circadian adrenal corticosterone rhythm following suprachiasmatic lesions in the rat. Brain Res 42:201–206.

Moore RY, Abrahamson EA, van den Pol A (2001) The hypocretin neuron system: an arousal system in the human brain. Arch Ital Biol 139:195–205.

Morison RS, Dempsey EW (1942) A study of thalamo-cortical relations. Am J Physiol 135:281–292.

Moruzzi G, Magoun HW (1949) Brain stem reticular formation and activation of the EEG. Electroencephalogr Clin Neurophysiol 1:455–473.

Mullaney DJ, Johnson LC, Naitoh P, Friedmann JK, Globus GG (1977) Sleep during and after gradual sleep reduction. Psychophysiology 14:237–244.

Munk MHJ, Roelfsema PR, König P, Engel AK, Singer W (1996) Role of reticular activation in the modulation of intracortical synchronization. Science 272:271–274.

Nagasaki H, Kitahama K, Valatx J-L, Jouvet M (1980) Sleep-promoting effect of the sleep-promoting substance (SPS) and delta sleep-inducing peptide (DSIP) in the mouse. Brain Res 192:276–280.

Nakanishi H, Sun Y, Nakamura RK, Mori K, Ito M, Suda S, Namba H, Storch FI, Dang TP, Mendelson W, Mishkin M, Kennedy C, Gillin JC, Smith CB, Sokoloff L (1997) Positive correlations between cerebral protein synthesis rates and deep sleep in *Macaca mulatta*. Eur J Neurosci 9:271–279.

Nauta WJH (1946) Hypothalamic regulation of sleep in rats. J Neurophysiol 9:285–316.

Nauta WJH, Kuypers HGJM (1958) Some ascending pathways in the brain stem reticular formation. In: Reticular formation of the brain (Jasper HH, Proctor LD, Knighton RS, Noshay WC, Costello RT, eds), pp 3–30. Boston: Little, Brown.

Nishino S, Ripley B, Overeem S, Lammers GJ, Mignot E (2000) Hypocretin (orexin) deficiency in human narcolepsy. Lancet 355:39–40.

Nofzinger EA, Mintun MA, Wiseman MB, Kupfer DJ, Moore RY (1997) Forebrain activation of REM sleep: an FDG PET study. Brain Res 770:192–201.

Norton S, de Beer EJ (1956) Effect of drugs on the behavioral patterns of cats. NY Acad Sci Ann 65:249–257.

O'Keefe J (1976) Place units in the hippocampus of the freely moving rat. Exp Neurol 51:78–109.

O'Keefe J, Dostrovsky J (1971) The hippocampus is a spatial map. Preliminary evidence from unit activity in the freely moving rat. Brain Res 34:171–175.

Oksenberg A, Shaffery JP, Marks GA, Speciale SG, Mihailoff G, Roffwarg HP (1996) Rapid eye movement sleep deprivation in kittens amplifies LGN cell-size disparity induced by monocular deprivation. Dev Brain Res 97:51–61.

Pappenheimer JR, Miller TB, Goodrich CA (1967) Sleep-promoting effects of cerebrospinal fluid from sleep-deprived goats. Proc Natl Acad Sci USA 58:513–518.

Pavlides C, Winson J (1989) Influences of hippocampal place cell firing in the awake state on the activity of these cells during subsequent sleep episodes. J Neurosci 9:2907–2918.

Pavlov IP (1923) The identity of inhibition with sleep and hypnosis. Sci Mon 17: 603–608.

Peñaloza–Rojas JH, Elterman M, Olmos N (1964) Sleep induced by cortical stimulation. Exp Neurol 10:140–147.

Percheron G, Mckenzie JS, Feger J, eds (1994) The basal ganglia, vol IV, New ideas on structure and function. New York: Plenum Press.

Peyron C, Tighe DK, van den Pol AN, de Lecca L, Heller HC, Sutcliffe JG, Kilduff TS (1998) Neurons containing hypocretin (orexin) project to multiple neuronal systems. J Neurosci 18:9996–10015.

Peyron C, Faraco J, Rogers W, Ripley B, Overeem S, Charnay Y, Nevsimalova S, Aldrich M, Reynolds D, Albin R, Li R, Hungs M, Pedrazzoli M, Padigaru M, Kucherlapati M, Fan J, Maki R, Lammers GJ, Bouras C, Kucherlapati R, Nishino S, Mignot E (2000) A mutation in a case of early onset narcolepsy and a generalized absence of hypocretin peptides in human narcoleptic brains. Nat Med 6:991–997.

Piéron H (1907) Le problème physiologique du sommeil. Paris: Masson.

Plihal W, Born J (1997) Effects of early and late nocturnal sleep on declarative and procedural memory. J Cogn Neurosci 9:534–547.

Plihal W, Born J (1999) Effects of early and late nocturnal sleep on priming and spatial memory. Psychophysiology 36:571–582.

Plumer SI, Siegel J (1973) Caudate-induced inhibition of hypothalamic attack behavior. Physiol Psychol 1:254–256.

Pompeiano O (1976) Mechanisms responsible for spinal inhibition during desynchronized sleep: experimental study. In: Advances in sleep research, vol 3, Narcolepsy (Guilleminault C, Dement WC, Passouant P, eds), pp 411–449. New York: Spectrum.

Pompeiano O, Pompeiano M, Corvaja N (1995) Effects of sleep deprivation on the post-natal development of visual-deprived cells in the cat's lateral geniculate nucleus. Arch Ital Biol 134:121–140.

Porkka-Heiskanen T, Strecker RE, Thakkar M, Bjørkum AA, Greene RW, McCarley RW (1997) Adenosine: a mediator of the sleep-inducing effects of prolonged wakefulness. Science 276:1265–1268.

Puolivāli J, Jākālā P, Koivisro E, Riekkinen Jr P (1998) Muscarinic M_1 and M_2 receptor subtype selective drugs modulate neocortical EEG via thalamus. NeuroReport 9:1685–1689.

Ramanathan L, Gulyani SA, Eiland MM, Gilliland M, Bergmann BM, Rechtschaffen A, Siegel JM (1999) Sleep deprivation induced neurodegenerative changes in the cortex of rats. Sleep Res Online 2 (suppl 1):551.

Ramm P, Smith CT (1990) Rates of cerebral protein synthesis are linked to slow-wave sleep in the rat. Physiol Behav 48:749–753.

Rechtschaffen A, Gilliland MA, Bergmann BM, Winter JB (1983) Physiological correlates of prolonged sleep deprivation in rats. Science 221:182–184.

Rechtschaffen A, Bergmann BM, Everson, CA, Kushida CA, Gilliland MA (1989) Sleep deprivation in the rat. X. Integration and discussion of the findings. Sleep 12:68–87.

Rechtschaffen A, Bergmann BM (2001) Letters to the editor. Am J Physiol Regul Integ Comp Physiol 280:R602–R603.

Rodriguez E, George N, Lachaux J-P, Martinerie J, Renault B, Varela FJ (1999) Perception's shadow: long-distance synchronization of human brain activity. Nature 397:430–433.

Roffwarg HP, Muzio J, Dement WC (1996) Ontogenetic development of the human sleep-dream cycle. Science 152:604–619.

Roldan E, Weiss T, Fifkova E (1963) Excitability changes during the sleep cycle of the rat. Electroencephalogr Clin Neurophysiol 15:775–785.

Ruckebusch Y (1962) Evolution post-natale du sommeil chez les ruminants. C R Soc Biol (Paris) 156:1869–1873.

Sacks O (1973) Awakenings. New York: HarperCollins.

Sakurai T, Amemiya A, Ishii M, Matsuzaki I, Chemelli RM, Tanaka H, Williams SC, Richardson JA, Kozlowski GP, Wilson S, Arch JRS, Buckingham RE, Haynes AC, Carr SA, Annan RS, McNulty DE, Liu W-S, Terrett JA, Elshourbagy NA, Bergsma DJ, Yanagisawa M (1998) Orexins and orexin receptors: a family of hypothalamic neuropeptides and G protein-coupled receptors that regulate feeding behavior. Cell 92:573–565; addendum p 696.

Sallanon M, Buda C, Janin M, Jouvet M (1985) Implication of serotonin in sleep mechanisms: induction, facilitation? In: Sleep: neurotransmitters and neuromodulators (Wauquier A, Gaillard JM, Monti JM, Radulovacki M, eds), pp 135–140. New York: Raven Press.

Sallanon M, Denayer M, Kitahama K, Aubert C, Gay N, Jouvet M (1989) Long-lasting insomnia induced by preoptic neuron lesions and its transient reversal by muscimol injection into the posterior hypothalamus in the rat. Neuroscience 32:669–683.

Saper CB (1985) Organization of cerebral cortical afferent systems in the rat. II. Hypothalamocortical projections. J Comp Neurol 237:21–46.

Saper CB (1987) Diffuse cortical projection systems: anatomical organization and role in cortical function. In: Handbook of physiology—the nervous system, vol V (Mountcastle VB, Plum F, eds), pp 169–210. Bethesda, MD: American Physiological Society.

Saper CB, Loewy AD (1980) Efferent projections of the parabrachial nucleus in the rat. Brain Res 197:291–317.

Saper CB, Sherin JE, Elmquist JK (1997) Role of the ventrolateral preoptic area in sleep induction. In: Sleep and arousal disorders: from molecule to behavior (Hayaishi O, Inoué S, eds), pp 281–294. Tokyo: Academic Press.

Sastre JP, Jouvet M (1979) Le comportement onirique du chat. Physiol Behav 22: 979–989.

Scheibel M, Scheibel A, Mollica A, Moruzzi G (1955) Convergence and interaction of afferent impulses on single units of reticular formation. J Neurophysiol 18:309–331.

Schlehuber CJ, Fleming DG, Lange GD, Spooner CE (1974) Paradoxical sleep in chickens. Behav Biol 11:537–546.

Schneider JS, Lidsky TI, eds (1987) Basal ganglia and behavior: sensory aspects of motor functioning. Toronto: Hans Huber.

Schoenberger GA, Monnier M (1977) Characterization of delta EEG sleep inducing peptide (DSIP). Proc Natl Acad Sci (Washington) 74:1282–1286.

Segundo JP, Arana R, French JD (1955a) Behavioral arousal by stimulation of the brain in the monkey. J Neurosurg 12:601–613.

Segundo JP, Naquet R, Buser P (1955b) Effect of cortical stimulation on electrocortical activity in monkeys. J Neurophysiol 18:236–245.

Shaw PJ, Cirelli C, Greenspan RJ, Tononi G (2000) Correlates of sleep and waking in *Drosophila melanogaster*. Science 287:1834–1837.

Shurley JT, Serafetinides EA, Brookes SE, Elsner R, Kenney DW (1969) Sleep in cetaceans. 1. The pilot whale, *Globicephala scammoni*. Psychophysiology 6:230.

Siegel J (1967) Recruiting response during states of sleep and arousal. Proceedings of the 75th Annual Convention of the American Psychological Association, pp 143–144.

Siegel J (1987) Striatal and limbic modulating influences: a search for target sites. In: Basal ganglia and behavior: sensory aspects of motor functioning (Schneider JS, Lidsky TI, eds), pp 183–195. Toronto: Hans Huber.

Siegel J, Brownstein RA (1975) Stimulation to N raphé dorsalis, central gray and hypothalamus: inhibitory and arousal effects. Physiol Behav 14:431–438.

Siegel J, Gordon T (1965) Paradoxical sleep deprivation in the cat. Science 148: 978–980.

Siegel J, Lineberry CG (1968) Caudate-capsular induced modulation of single unit activity in mesencephalic reticular formation. Exp Neurol 22:444–463.

Siegel J, Wang RY (1974) Electroencephalographic, behavioral, and single-unit effects produced by stimulation of forebrain inhibitory structures in cats. Exp Neurol 42:28–50.

Siegel JM (1985) Ponto-medullary interactions in the generation of REM sleep. In: Brain mechanisms of sleep (McGinty DJ, Drucker-Colin R, Morrison A, Parmeggiani PL, eds), pp 157–174. New York: Raven Press.

Siegel JM (2000a) Narcolepsy. Sci Am 282:76–81.

Siegel JM (2000b) Brain stem mechanisms generating REM sleep. In: Principles and practice of sleep medicine, 3rd ed (Kryger MH, Roth T, Dement WC, eds), pp 112–133. Philadelphia: WB Saunders.

Siegel JM, Manger PR, Nienhuis R, Fahringer HM, Pettigrew JD (1996) The echidna *Tachyglossus aculeatus* combines REM and non-REM aspects in a single sleep state: implication for the evolution of sleep. J Neurosci 16:3500–3506.

Siegel JM, Manger PR, Nienhuis R, Fahringer HM, Shalita T, Pettigrew JD (1999) Sleep in the platypus. Neurosci 91:391–400.

Siegel JM, Tomaszewski KS (1983) Behavioral organization of reticular formation: studies in the unrestrained cat. I. Cells related to axial, limb, eye, and other movements. J Neurophysiol 50:696–716.

Siegel JM, Wheeler RL, McGinty DJ (1979) Activity of medullary reticular formation neurons in unrestrained cat during waking and sleep. Brain Res 179:49–60.

Silver R, LeSauter J, Tresco PA, Lehman M (1996) A diffusible coupling signal from the transplanted suprachiasmatic nucleus controlling circadian locomotive rhythms. Nature 382:810–813.

Singer W (1993) Synchronization of cortical activity and its putative role in information processing and learning. Annu Rev Physiol 55:349–374.

Skadberg BT, Markestad T (1997) Consequences of getting the head covered during sleep in infancy. Pediatrics (electronic pages) 100:e6, 1–7.

Smith C (1995) Sleep states and memory processes. Behav Brain Res 69:137–145.

Smith C (1996) Sleep states, memory processes and synaptic plasticity. Behav Brain Res 78:49–56.

Smith C (2001) Sleep states and memory processes in humans: procedural versus declarative memory systems. Sleep Med Rev 5:491–506.

Starzl TE, Taylor CW, Magoun HW (1951) Ascending conduction in reticular activating system, with special reference to the diencephalon. J Neurophysiol 14: 461–477.

Steinschneider A (1972) Prolonged apnea and the sudden infant death syndrome: clinical and laboratory observations. Pediatrics 50:646–654.

Stephan FK, Zucker I (1972) Circadian rhythms in drinking behavior and locomotor activity are eliminated by hypothalamic lesions. Proc Natl Acad Sci USA 69: 1583–1586.

Steriade M (1996) Arousal: revisiting the reticular activating system. Science 272: 225–226.

Steriade M, Amzica F, Contreras D (1996a) Synchronization of fast (30–40 Hz) spontaneous cortical rhythms during brain activation. J Neurosci 16:392–417.

Steriade M, Contreras D, Amzica F, Tomofeev I (1996b) Synchronization of fast (30–40 Hz) spontaneous oscillations in intrathalamic and thalamocortical networks. J Neurosci 16:2788–2808.

Steriade M, Datta S, Pare D, Oakson G, Currodossi R (1990) Neuronal activities in brain-stem cholinergic nuclei related to tonic activation processes in thalamocortical systems. J Neurosci 10:2541–2559.

Steriade M, Deschênes M (1984) The thalamus as a neuronal oscillator. Brain Res Rev 8:1–63.

Steriade M, McCarley RW (1990) Brainstem control of wakefulness and sleep. New York: Plenum Press.

Sterman MB, Clemente CD (1962a) Forebrain inhibitory mechanisms: cortical synchronization induced by basal forebrain stimulation. Exp Neurol 6:91–102.

Sterman MB, Clemente CD (1962b) Forebrain inhibitory mechanisms: sleep patterns induced by basal forebrain stimulation in the behaving cat. Exp Neurol 6:103–117.

Stickgold R, LaTanya J, Hobson JA (2000a) Visual discrimination learning requires sleep after training. Nat Neurosci 3:1237–1238.

Stickgold R, Whidbee D, Schirmer B, Patel V, Hobson JA (2000b) Visual discrimination task improvement: a multi-step process occurring during sleep. J Cogn Neurosci 12:246–254.

Szymusiak R, McGinty D (1986) Sleep suppression following kainic acid–induced lesions of the basal forebrain. Exp Neurol 94:598–614.

Thakkar MM, Strecker RE, Delgiacco RA, McCarley RW (1999) Adenosinergic A_1 inhibition of basal forebrain wake-active neurons: a combined unit recording and microdialysis study in freely behaving cats. Sleep Res Online 2 (suppl 1):91–92.

Thannickal TC, Moore RY, Nienhuis R, Ramanathan L, Gulyani S, Aldrich M, Cornford M, Siegel JM (2000) Reduced number of hypocretin neurons in human narcolepsy. Neuron 27:469–474.

Thorpy MJ (2000) Classification of sleep disorders. In: Principles and practice of sleep medicine, 3rd ed (Kryger MH, Roth T, Dement WC, eds), pp 547–557. Philadelphia: WB Saunders.

Villablanca J (1965) The electrocorticogram in the chronic cerveau isolé cat. Electroencephalogr Clin Neurophysiol 19:576–586.

Vimont-Vicary P, Jouvet-Mounier D, Delorme F (1966) Effets EEG et comportement des privations de sommeil paradoxal chez le chat. Electroencephalogr Clin Neurophysiol 20:439–449.

Vogel GW (1979) A motivational function of REM sleep. In: The functions of sleep (Drucker-Colin R, Shkurovich M, Sterman MB, eds), pp 233–250. New York: Academic Press.

Vogel GW, Vogel F, McAbee RS, Thurmond AJ (1980) Improvement of depression by REM sleep deprivation. Arch Gen Psychiat 37:247–253.

Von Economo C (1923) Encephalitis lethargica. Wien Med Wschr 73:777–782.

Von Economo C (1930) Sleep as a problem of localization. J Nerv Ment Dis 71:248–259.

Wasman M, Flynn JP (1962) Directed attack elicited from hypothalamus. Arch Neurol 6:220–227.

Webb WB (1992) Sleep, the gentle tyrant. Bolton, MA: Anker.

Welsh DK, Logothetis DE, Meister M, Reppert SM (1995) Individual neurons dissociated from rat suprachiasmatic nucleus express independently phased circadian firing rhythms. Neuron 14:697–706.

Wilson MA, McNaughton BL (1994) Reactivation of hippocampal ensemble memories during sleep. Science 265:676–679.

Xi M-C, Morales FR, Chase MH (1999) A GABAergic reticular system is involved in the control of wakefulness and sleep. Sleep Res Online 2:43–48.

Zhang J, Obál Jr F, Zheng T, Fang J, Taishi P, Krueger JM (1999) Intrapreoptic microinjection of GHRH or its antagonist alters sleep in rats. J Neurosci 19:2187–2194.

Zheng D, Purves D (1995) Effects of increased neural activity on brain growth. Proc Natl Acad Sci USA 92:1802–1806.

Abbreviations

APSS	Association for the Psychophysiological Study of Sleep
ARAS	ascending reticular activating system
ASDA	American Sleep Disorders Association
ATP	adenosine triphosphate
canarc-1	canine narcolepsy gene
CPAP	continuous positive airway pressure
CR	conditioned response
CS	conditioned stimulus
CSF	cerebrospinal fluid
DCSD	Diagnostic Classification of Sleep and Arousal Disorders
DSIP	delta-sleep-inducing peptide
EEG	electroencephalogram, electroencephalograph, electroencephalography
EMG	electromyogram
EOG	electro-oculogram
EPSP	excitatory postsynaptic potential
GABA	γ-aminobutyric acid
GHRH	growth hormone releasing hormone
HLA	human leukocyte antigen
ICSD	International Classification of Sleep Disorders
IL-1	interleukin 1
IPSP	inhibitory postsynaptic potential
LDT	laterodorsal tegmental [nucleus]
LVFA	low voltage, fast activity
LVHF	low voltage, high frequency
MRF	mesencephalic reticular formation
NPO	nucleus pontis oralis
NREM	non–rapid eye movement [sleep]
NTS	nucleus of the tractus solitarius
OSA	obstructive sleep apnea
PCPA	p-chlorophenylalanine

PET	positron emission tomography
PGO	ponto-geniculo-occipital [spikes]
PPT	pedunculopontine tegmental [nucleus]
RAS	reticular activating system
REM	rapid eye movement
RF	reticular formation
SCN	suprachiasmatic nucleus
SIDS	sudden infant death syndrome
SPS	sleep-promoting substance
SWS	slow-wave sleep
UR	unconditioned response
US	unconditioned stimulus

Glossary

acetylcholine   See **cholinergic neurons**.

action potential   The all-or-none electrical signal generated by neurons that is conducted along the axon and causes the release of neurotransmitters from axon terminals. See also **compound action potential**.

activated sleep   Another term for REM sleep, often used with reference to REM sleep in animals. See **REM sleep**.

adenosine   A product derived from the breakdown of adenosine triphosphate (ATP) during metabolic processes. Adenosine increases in the brain when neural activity increases, as during waking.

agonist   A drug that facilitates or mimics the effects of a neurotransmitter.

aminergic neurons   Neurons that synthesize neurotransmitters called the monoamines or biogenic amines. These include norepinephrine, epinephrine, dopamine, serotonin, and histamine. See also **norepinephrine, epinephrine, dopamine, serotonin,** and **histamine**.

antagonist   A drug that opposes the action of a neurotransmitter.

antidromic response   An action potential that is artificially generated at axon terminals or in the axon and conducted back to the cell body where it is recorded.

alpha waves   A regular EEG rhythm in the frequency range of 8 to 10 Hz. It usually occurs during periods of quiet relaxation. See also **Berger rhythm**.

ascending reticular activating system (ARAS)   A region of the mesencephalic and pontine reticular formation that, when stimulated, causes cortical arousal. See also **reticular formation**.

autoimmune response   A response in which the destructive force of the immune system is directed against the organism's own tissue. Cells of specific types are attacked as if they were foreign intruders (e.g., invading bacteria).

autoreceptors   Receptors on a neuron that are sensitive to the neurotransmitter released by that neuron.

axoplasmic transport   A property of neurons in which material is transported down the axon from the cell body to the terminals (anterograde transport) and from the terminals back to the cell body (retrograde). This property of neurons has been used for tracing neuroanatomical pathways.

baroceptors   Pressure receptors located in the carotid sinus that respond to changes in blood pressure. These receptors convey their blood pressure–induced signals to the brain stem. See also **carotid sinus.**

basal forebrain   The ventral part of the telencephalon that is rostral to and continuous with the preoptic area of the anterior hypothalamus. See also **telencephalon** and **hypothalamus.**

basal ganglia   Subcortical nuclei of the telencephalic forebrain that are involved in the control of movement. See also **forebrain** and **caudate nucleus.**

Berger rhythm   An older term that refers to the alpha waves. Named after Hans Berger, the "father" of electroencephalography. See also **alpha waves.**

beta waves   An EEG rhythm in the frequency range of 20 to 40 Hz, associated with states of cortical activation seen in waking and in REM sleep.

binding phenomenon   The integration (binding) of neural activity in different parts of the brain to create a functional circuit. This is considered to be the basis of unified perceptions that involve the neural activity in anatomically disparate regions of the brain. See **gamma rhythm.**

biogenic amines   See **aminergic neurons.**

blood–brain barrier   The semipermeable barrier between brain cells and the blood supply to the brain that prevents certain blood-borne substances from affecting the brain.

brachium conjunctivum   A fiber bundle from the cerebellum to the forebrain that passes through the pontine and mesencephalic reticular formation. It is also called the superior cerebellar peduncle.

brain stem   Part of the brain that is continuous with the spinal cord at the point where the spinal cord enters the skull. The brain stem comprises the medulla oblongata, pons, and mesencephalon. Some anatomists include the diencephalon as the most rostral extent of the brain stem. See also **medulla oblongata, pons, mesencephalon,** and **diencephalon.**

canarc-1   The gene for canine narcolepsy. See also **narcolepsy.**

carotid sinus   An enlargement of the carotid arteries located in the neck that contain baroceptors. See also **baroceptors.**

cataplexy   A sudden episode of muscle weakness or a profound loss of muscle tonus (paralysis) during waking. A symptom of narcolepsy. See also **narcolepsy** and **muscle atonia.**

caudate nucleus   A nucleus of the basal ganglia that has functions related to movement and behavioral inhibition.

cerebellar peduncles   Three bilateral pairs of fiber tracts that connect the cerebellum to the brain stem and the rest of the brain. They are the superior, middle, and inferior cerebellar peduncles. See also **cerebellum** and **brachium conjunctivum.**

cerebellum A structure attached to the brain stem at the pontine level by the cerebellar peduncles. It is involved with motor coordination, balance, and posture. See also **cerebellar peduncles**.

cerebral cortex Layers of cell bodies with their dendritic fields and associated axons that cover the entire cerebrum. See also **cerebrum**.

cerebral hemispheres See **cerebrum**.

cerebral ventricles Hollow spaces within the cerebrum filled with cerebrospinal fluid. See also **cerebrum** and **cerebrospinal fluid**.

cerebrospinal fluid (CSF) A clear fluid similar to the plasma component of blood. It is found in the four interconnected cerebral ventricles and flows from there to an enclosed space (the subarachnoid space) that covers the entire brain and spinal cord. See also **cerebral ventricles**.

cerebrum The telencephalic portion of the forebrain comprised of the two large cerebral hemispheres. See also **forebrain**.

cerveau isolé A surgical preparation in which the brain is transected between the diencephalon and mesencephalon. See also **diencephalon** and **mesencephalon**.

cholinergic neurons Neurons that release the neurotransmitter acetylcholine at their terminals.

circadian rhythm A physiological or behavioral function that occurs and repeats itself on a daily basis.

cisterna magna A membrane-bound reservoir that collects the outflow of cerebrospinal fluid from the fourth ventricle. Located at the base of the brain at the junction of the spinal cord, medulla, and cerebellum. See also **cerebral ventricles** and **cerebrospinal fluid**.

compound action potential The neural activity recorded from a fiber bundle or whole nerve. This response to a discrete stimulus is the sum of the individual action potentials that occur in near simultaneity. See also **action potential**.

corticospinal fibers Fibers that originate in motor cortex and project to the spinal cord.

cytokines Protein molecules released by the immune system in response to invading microorganisms. Cytokines produce, among other things, fever and sleep.

declarative memory The recall of episodes of which one is consciously aware. Also called episodic or explicit memory.

delayed conditioning The classical (Pavlovian) conditioning paradigm in which there is a delay (e.g., one minute) between the onset of the conditioned stimulus (CS) and the unconditioned stimulus (US).

delta waves An EEG rhythm in the frequency range of 1 to 3 Hz, associated with deep slow-wave, stage 4, sleep. See also **slow-wave sleep** and **stage 4 sleep**.

dendritic field potential The summated electrical activity recorded from the dendritic fields of a population of cells; the basis of the usual EEG recordings. This neural activity in response to a discrete stimulus is the basis of the evoked (field) potential.

desynchronized sleep   See **REM sleep.**

desynchrony   An EEG pattern characterized by low voltage, fast activity waves seen during waking and REM sleep.

diencephalon   The caudal portion of the forebrain located between the telencophalon, rostrally, and the mesencephalon, caudally. See also **forebrain.**

dopamine   See **dopaminergic neurons.**

dopaminergic neurons   Neurons that release the neurotransmitter dopamine at their terminals.

dream sleep   See **REM sleep.**

dyssomnias   A category of sleep disorders characterized by difficulty in falling asleep and remaining asleep. Ordinarily called insomnia.

EEG desynchronization   A recording (usually from the cerebral cortex) of low voltage fast activity (LVFA) waves indicative of a state of waking or of REM sleep. See also **electroencephalogram, low voltage, fast activity (LVFA) waves,** and **REM sleep.**

EEG synchrony   A recording (usually from the cerebral cortex) of high voltage, low frequency (HVLF) waves indicative of a state of slow-wave sleep (SWS). See also **electroencephalogram, high voltage, low frequency (HVLF) waves,** and **slow-wave sleep (SWS).**

electroencephalogram (EEG)   An electrical record of brain wave activity.

electroencephalograph (EEG)   The machine that records brain wave activity.

electromyogram (EMG)   An electrical record of muscle activity.

electro-oculogram (EOG)   An electrical record of eye movements.

encéphale isolé   A surgical preparation in which the brain is separated from the spinal cord by a transection between the medulla and spinal cord. See also **medulla oblongata.**

encephalitis lethargica   A brain disease characterized by constant sleep. This is caused by the destruction of cells in the reticular formation and posterior hypothalamus that are necessary for wakefulness. See also **reticular formation** and **hypothalamus.**

encephalization   The evolutionary process characterized by the forebrain becoming larger, more complex, and gaining some degree of control over lower parts of the brain. See also **forebrain.**

episodic memory   See **declarative memory.**

evoked field potential   See **dendritic field potential.**

excitatory postsynaptic potential (EPSP)   A brief depolarization of dendritic or somatic membrane by action of a neurotransmitter. The summation of this potential with other EPSPs increases the likelihood that an action potential will be generated in the postsynaptic neuron. See also **action potential.**

explicit memory   See **declarative memory**.

extrathalamic projection nuclei   Subcortical nuclei that project directly to the cortex without first synapsing in the thalamus. See also **thalamus**.

field potential   See **dendritic field potential**.

forebrain   The most rostral part of the brain, comprising the telencephalon (cerebral hemispheres) and the diencephalon. Also called the prosencephalon. See also **telencephalon** and **diencephalon**.

GABA (γ-aminobutyric acid)   An inhibitory neurotransmitter.

GABAergic neurons   Neurons that release the neurotransmitter GABA at their terminals. See also **GABA**.

galvanometer   A voltage-measuring device.

gamma rhythm   A low voltage EEG rhythm in the frequency range of 30 to 60 Hz (often called a 40 Hz rhythm), associated with the binding phenomenon. See also **binding phenomenon**.

gliosis   A proliferation of a type of glial cells as a reaction to nerve cell damage and nerve death. Gliosis is a form of neural scar tissue.

glutamate   An excitatory neurotransmitter. See also **glutamatergic neurons**.

glutamatergic neurons   Neurons that release the neurotransmitter glutamate at their terminals. See also **glutamate**.

glycine   An inhibitory neurotransmitter.

graded potentials   Voltages (potentials) that occur with varying amplitudes (graded). Postsynaptic potentials (EPSPs and IPSPs) and dendritic field potentials are graded, in contrast to single nerve cell action potentials, which have all-or-none amplitude characteristics. See also **dendritic field potential, action potential, EPSP,** and **IPSP**.

growth hormone releasing hormone (GHRH)   A hormone produced by cells of the hypothalamus that causes the release of the growth hormone from the pituitary gland. Another group of hypothalamic neurons produces GHRH as a neurotransmitter that has a sleep-inducing effect.

high voltage, low frequency (HVLF) waves   EEG rhythms comprising of sleep spindles and delta waves seen during sleep stages 3 and 4. See also **slow-wave sleep (SWS), sleep spindles,** and **delta waves**.

hindbrain   The most caudal part of the brain stem, comprising the pons and medulla oblongata. Also called the rhombencephalon. See also **pons** and **medulla oblongata**.

hippocampus   A structure in the telencephalic forebrain involved in emotional behavior and memory function.

histamine   See **histaminergic neurons**.

histaminergic neurons   Neurons that release the neurotransmitter histamine at their terminals. See also **tuberomammillary nucleus.**

human leukocyte antigens (HLA)   A family of genes that, when expressed, trigger an autoimmune response. See also **autoimmune response.**

hypersomnia   An increased amount of sleep.

hypnogenic   A substance or manipulation that induces sleep.

hypnogogic hallucination   Imaginary, dreamlike perceptions that occur during waking or the transition between sleep and waking.

hypnogram   A graphic representation of the stages of sleep over the course of a night. It summarizes the durations and timing of the sleep stages in histogram form.

hypnotoxin   A substance that induces sleep.

hypocretin   A peptide neurotransmitter localized to a group of cells in the hypothalamus. Also called orexin.

hypopnea   A decrease in the respiration rate

hyposomnia   A reduced amount of sleep.

hypothalamus   A group of nuclei located in the ventral diencephalon and involved in the neural control of autonomic or visceral processes. Certain nuclei control circadian rhythms and other functions related to sleep and waking. See also **diencephalon.**

immune response   Physiological processes used by the body to combat invading microorganisms as well as other foreign materials. See also **cytokines, human leukocyte antigens (HLA), muramyl peptides,** and **autoimmune response.**

implicit memory   See **procedural memory.**

inhibitory postsynaptic potential (IPSP)   A brief hyperpolarization (a graded potential) of the dendrites or soma of a neuron by the synaptic action of a neurotransmitter. This renders the neuron less likely to generate an action potential. See also **graded potential** and **action potential.**

intergeniculate leaflet   Part of the lateral geniculate nucleus of the thalamus that provides light information to the suprachiasmatic nucleus of the hypothalamus. This influence participates in the control of circadian rhythms. See also **thalamus, suprachiasmatic nucleus (SCN),** and **circadian rhythm.**

internal capsule   A large fiber bundle in the telencephalon with ascending and descending components that connect the cerebral cortex with regions below the cortex. See also **cerebral cortex.**

intralaminar nuclei   Thalamic nuclei that project to broad areas of the cerebral cortex. They are sometimes referred to as "nonspecific" or "diffuse" projection nuclei. See also **thalamus.**

K-complex   A sharp, high voltage transient wave seen during stage 2 sleep that occurs spontaneously and intermittently or may be triggered by a sensory stimulus. See also **stage 2 sleep.**

lateral geniculate nucleus A nucleus of the thalamus that receives visual information from the retina, processes it, and relays it to the visual cortex. See also **thalamus**.

laterodorsal tegmental (LDT) nucleus One of the peribrachial nuclei of the reticular activating system. See **peribrachial nuclei** and **ascending reticular activating system (ARAS)**.

ligand A chemical that binds to a protein, such as a postsynaptic receptor. Neurotransmitters are natural ligands that bind to receptor sites.

limbic system Brain structures, many of which are in the forebrain, that participate in the control of emotional behaviors. See also **forebrain**.

locus coeruleus A nucleus in the dorsal pons that is the major source of norepinephrine cells in the brain. See also **pons**.

low voltage, fast activity (LVFA) waves An EEG rhythm comprising beta waves seen during waking and REM sleep. Also called low voltage, high frequency waves. See also **beta waves** and **REM sleep**.

low voltage, high frequency (LVHF) waves See **low voltage, fast activity waves**.

mandibular advancement appliance A dental appliance attached to the lower jaw (mandible) that pulls the jaw forward to facilitate keeping the airway open. Used for preventing sleep apnea. See also **sleep apnea**.

marsupial mammals The order of mammals that carry their immature young in pouches.

medulla oblongata The most caudal part of the brain stem. See also **brain stem**.

melatonin A hormone secreted by the pineal gland that is involved in the control of seasonal breeding cycles and daily sleep and waking cycles. See also **pineal gland**.

memory consolidation The neural events in the brain responsible for memory storage.

mesencephalon The part of the brain stem located between the forebrain and hindbrain. Also called the midbrain. See also **brain stem, forebrain, and hindbrain**.

microsleep A period during which a person falls asleep briefly (on the order of seconds). Usually occurs in sleep-deprived individuals during boring and tedious tasks.

midbrain See **mesencephalon**.

midpontine pretrigeminal section A transection of the brain stem at the midpontine level just rostral to the sensory and motor roots of the trigeminal nerve. See also **brain stem** and **pons**.

monotremes The most primitive order of mammals. They secrete milk, but also lay eggs.

muramyl peptides Protein residues of killed and digested bacteria that stimulate the immune system and produce sleep.

muscimol   A GABA agonist. See also **GABA** and **agonist.**

muscle atonia   A decrease or loss of muscle tension seen during REM sleep and cataplexy. See also **REM sleep** and **cataplexy.**

narcolepsy   A sleep disorder characterized by excessive daytime sleepiness that often culminates in sudden sleep attacks. See also **cataplexy.**

noradrenaline   Another name for norepinephrine. See **noradrenergic neurons.**

noradrenergic neurons   Neurons that release the neurotransmitter norepinephrine (noradrenaline) at their terminals. See also **noradrenaline** and **norepinephrine.**

norepinephrine   Another name for noradrenaline. See **noradrenergic neurons.**

nucleus of the tractus solitarius (NTS)   A nucleus located in the pons. Stimulation of a region of this nucleus will produce EEG cortical synchronization. See also **pons** and **EEG synchrony.**

nucleus pontis oralis (NPO)   A cell group in the rostral pons that has an excitatory influence on cholinergic cells of the basal forebrain (BFB). Activation of the BFB, in turn, produces the EEG cortical activation seen in REM sleep. See also **pons, cholinergic neurons,** and **basal forebrain.**

nucleus subcoeruleus   Located in the pontine reticular formation. A lesion of this nucleus produces a loss of muscle atonia during REM sleep (i.e., muscle tonus remains during REM sleep). See also **REM sleep.**

ontogeny   The developmental history of the individual organism.

orexin   A peptide neurotransmitter also known as hypocretin. See **hypocretin.**

orthodromic activity   An action potential that is conducted down an axon in the normal direction, (i.e., from the cell body to the axon terminals). See also **action potential.**

paradoxical sleep   Another term for REM sleep, usually used for animals. See **REM sleep.**

parasomnias   Disorders of arousal, REM sleep, and the transition state between sleep and waking. Examples are sleepwalking and sleeptalking, nightmares, and sleep paralysis. See also **REM sleep** and **sleep paralysis.**

pedunculopontine tegmental (PPT) nucleus   One of the peribrachial nuclei of the reticular activating system. See **peribrachial nuclei** and **ascending reticular activating system (ARAS).**

penetrance   The degree of probability of a gene being expressed: high penetrance, high probability.

peribrachial nuclei   The peribrachial nuclei are the pedunculopontine tegmental (PPT) nucleus and the laterodorsal tegmental (LDT) nucleus located at the pontomesencephalic junction. The cholinergic cells of these nuclei are a major component of the reticular activating system responsible for the cortical arousal seen in waking and in REM sleep. See also **ascending reticular activating system (ARAS), cholinergic neurons, pons,** and **mesencephalon.**

pharynx   The rear part of the oral cavity, where the opening to the trachea (airway) is located. See also **trachea.**

phylogeny   The evolutionary history of a species or other taxonomic category.

pineal gland   A gland located at the dorsal extent of the diencephalon that secretes melatonin. See also **melatonin** and **diencephalon.**

pituitary gland   A major endocrine gland located just below the hypothalamus. See also **hypothalamus.**

placental mammals   mammalian species that are protected and nourished by the placenta in the uterus prior to birth and nourished by milk after birth.

pons   The rostral portion of the hindbrain located betwen the medulla oblongata, caudally, and the mesencephalon, rostrally.

ponto-geniculo-occipital (PGO) spikes   Sharp transients in the EEG recorded from the pons, lateral geniculate nucleus of the thalamus, and the occipital (visual) cortex. They occur intermittently during REM sleep. See also **electroencephalogram (EEG), pons, lateral geniculate nucleus,** and **REM sleep.**

preoptic area   The most rostral part of the hypothalamus that is continuous with the basal forebrain area. See also **hypothalamus** and **basal forebrain.**

procedural memory   The learning and retention of certain skills—like those involved in motor learning. One is not consciously aware of the process by which these skills are acquired. Also called implicit memory.

prone   Lying with the face and belly down.

prosencephalon   See **forebrain.**

pyramidal tract   A fiber tract that originates in motor cortex that carries messages to motor nuclei of cranial nerves and to motor neurons of the spinal cord. See also **corticospinal fibers.**

raphé nuclei   Nuclei located along the midline of the entire brain stem. Raphé neurons synthesize serotonin as their neurotransmitter. See also **serotonergic neurons.**

rapid eye movement (REM) sleep   The stage of sleep during which dreaming occurs. (Dreams have also been reported during other stages of sleep.) This sleep stage is characterized by a desynchronized EEG record, rapid eye movements (REMs), loss of muscle tonus (atonia), and PGO spikes. See also **EEG desynchronization, muscle atonia,** and **PGO spikes.**

receptor antagonist   A drug that blocks the action of a neurotransmitter at the receptor site for the transmitter.

recruiting waves   A form of EEG synchronization that resembles sleep spindles. Elicited by low frequency stimulation of the intralaminar nuclei of the thalamus. See also **EEG synchronization, sleep spindles, intralaminar nuclei,** and **thalamus.**

reticular formation   A network of neuronal cell bodies interspersed among nerve fibers located throughout the brain stem tegmentum from the medulla to the mesencephalon. It contains many nuclei, some of which are important for the regulation of sleep and waking. See also **brain stem** and **tegmentum.**

reticular nucleus of the thalamus   This nucleus plays a major role in regulating the arousal state of the cortex via its control function over other nuclei of the thalamus. See also **thalamus.**

retinohypothalamic tract   A component of the optic tract that projects form the retina to the suprachiasmatic nucleus (SCN) of the hypothalamus. It provides information about light and darkness for the entrainment of certain circadian rhythms. See **circadian rhythm** and **suprachiasmatic nucleus (SCN).**

rhombencephalic sleep   Another term for REM sleep. See **REM sleep.**

rhombencephalon   See **hindbrain.**

sagittal section   A vertical section that extends from the front to the back of the brain. A medial or midsagittal section divides the brain into left and right halves. Other sagittal sections are parallel to that.

serotonergic neurons   Neurons of raphé nuclei that release the neurotransmitter serotonin at their terminals.

serotonin   See **serotonergic neurons.**

sleep apnea   A cessation of breathing during sleep that lasts 10 seconds or more.

sleep paralysis   Loss of muscle tonus and of voluntary movement that some people experience during a brief transition state between sleep and waking.

sleep spindles   An EEG waveform having a frequency of 12 to 15 Hz that waxes and wanes in amplitude, with the highest waves in the middle of a burst 1 to 2 seconds long. The envelope of this waveform is spindle shaped and occurs during stages 2 and 3 of sleep. See also **high voltage, low frequency (HVLF) waves, slow-wave sleep, stage 2 sleep,** and **stage 3 sleep.**

slow-wave sleep (SWS)   Stages 3 and 4 of sleep, characterized by sleep spindles and delta waves. See also **sleep spindles, delta waves, high voltage, low frequency (HVLF) waves, stage 3 sleep,** and **stage 4 sleep.**

somnogenic   See **hypnogenic.**

stage 1 sleep   A sleep stage characterized by a low voltage, high frequency EEG, similar to beta waves seen in waking and in REM sleep. Also called stage 1 non-REM (NREM) sleep. Considered by some to be a transition state between waking and sleep. See also **low voltage, high frequency (LVHF) waves** and **beta waves.**

stage 2 sleep   A sleep stage characterized by sleep spindles interspersed between low voltage, high frequency (LVHF) waves. See also **sleep spindles** and **low voltage, high frequency (LVHF) waves.**

stage 3 sleep   A sleep stage characterized by sleep spindles interspersed between delta waves. See also **sleep spindles** and **delta waves.**

stage 4 sleep   The deepest stage of non-REM sleep, characterized by delta waves. See also **delta waves.**

stereotaxic instrument   A device used in brain surgery to hold the head rigid and in a fixed, standard orientation. This device permits the accurate placement of probes to manipulate (stimulate, record, lesion) regions within the brain.

superior cerebellar peduncle   See **brachium conjunctivum.**

supine   Lying on the back with the face up.

suprachiasmatic nucleus (SCN)   Located in the hypothalamus. The control nucleus for the circadian clock. See also **circadian rhythm** and **hypothalamus.**

synchronized sleep   Sleep characterized by EEG slow waves, namely, sleep spindles and delta waves. See also **slow-wave sleep (SWS), high voltage, low frequency (HVLF) waves, sleep spindles,** and **delta waves.**

tegmentum   The central core of the brain stem that contains the reticular formation and numerous brain stem nuclei and fiber tracts. See also **brain stem** and **reticular formation.**

telencephalon   The rostral part of the forebrain. See also **cerebrum.**

thalamus   A complex of nuclei located in the dorsal part of the diencephalon. They comprise the main input to the cerebral cortex. The various nuclei of the thalamus are involved in sensory, motor, emotional, and arousal functions. See also **diencephalon, lateral geniculate nucleus, intralaminar nuclei,** and **reticular nucleus of the thalamus.**

theta waves   An EEG rhythm at 4 to 8 Hz, most often at 6 Hz.

trachea   The airway passage leading from its opening in the pharynx to the lungs. See also **pharynx, sleep apnea,** and **mandibular advancement appliance.**

tuberomammillary nucleus   A nucleus of the posterior hypothalamus, the cells of which synthesize histamine as their neurotransmitter. See also **histaminergic neurons.**

ventricles   See **cerebral ventricles.**

Index